Principles of Growth and Development

Principles of Growth and Development

Editors

Monica Tomar
S.K. Sharma

ISHA BOOKS
DELHI-110 033

Principles of Growth and Development

Rs. 690

ISBN: 81-8205-155-X

Published in 2005 in India by
Isha Books, D-43, Prithviraj Road,
Adarsh Nagar, Delhi-110 033

Laser Typesetting at: Shree Vaishnavi Creations, Delhi
Printed at: Mehra Offset Press, Delhi

PREFACE

There is a set of principles that characterises the pattern and process of growth and development. These principles or characteristics describe typical development as a predictable and orderly process; that is, we can predict how most children will develop and that they will develop at the same rate and at about the same time as other children. Although there are individual differences in children's personalities, activity levels, and timing of developmental milestones, such as ages and stages, the principles and characteristics of development are universal patterns.

Child development involves complex changes in related processes and behaviours; it is not a product of isolated transformations. Physical, cognitive, social, and emotional development are intimately interrelated. Infants, children, and adolescents are influenced not only by personal relationships, but also by social policies and their cultural environments. Also, children play an active role in their own development: their characteristics and actions cause others to react in certain ways, and these reactions then influence their development. This book focuses on the influences these intertwined interactions have on the physical, cognitive, social, and emotional development of infants, children, and adolescents.

The best way to study and understand children and their development is to consider development within different biological, familial, social, and cultural backgrounds. The dynamic interplay among diverse individual and environmental characteristics means that children's developmental outcomes are influenced by many factors. An understanding of the principles of development helps us to plan appropriate activities and stimulating and enriching experiences for children, and provides a basis for understanding how to encourage and support young children's learning.

This book explores the principles of physical, mental and emotional growth and development in children and teenagers. It provides information on the activities and achievements usually displayed at a given stage of development and also direct readers to information on how to help children and adolescents reach their full potential as they grow and develop. To students considering employment in teaching, nursing, psychology, day-care centres, etc., this book is destined to become a valuable reference tool.

Editor

CONTENTS

1

Concept of Growth and Development

The concept of human development in the broadest sense encompasses virtually anything that affects humans. Human development theory includes the biological, social and psychological changes that occur as humans growth and develop.

STAGE AND NON-STAGE THEORY

Stage theorists assume that growth and development happens in terms of definable stages generally related to age. Development differs from stage to stage which varies across different theories. Thus, Piaget might say that a four year old child can perform certain mental functions. Obviously, there are enough individual differences to dispute much of the thinking of the stage theorists. Some characteristic proponents have been Freud, Piaget, Kohlberg and Erikson. Non-Stage theorists view growth and development as a continuous process. Such theorists do not categorize development into stages i.e. not broken into definable stages.

Generally such theorists stress the environment as the major influence. That same four year old might be able to perform certain mental functions that some cannot perform until age six. The adult of forty may be locked into Loevinger's stage four or five while another adult of thirty may be functioning at stage six. Some characteristic proponents are Loevinger and Bandura. While there are general patterns that will apply to the majority of the population, there is not a pattern that fits all, nor is there a theory that predicts human development for all. Experiences and how different people process those experiences can have a great impact on how we grow and develop. One thing we know for sure is that almost all of us continue to develop in some way through much of out lives, including our adult life.

Some categorize theories into different groupings. The developmental task group includes life age theorists and life cycle theorists (Erikson, Havighurst, Neugarten) And developmental stage theorists (Kohlberg, Piaget, Loevinger, Harvey, Hunt, Schroeder and Gregoric)

Categories of theories

Developmental theories

1. Biological and psychological changes that occur in people during various times in their lives.
2. Apply to all people with similar biological, physical and social backgrounds.
3. Tend to be broader than for example, learning theory or teaching theory because these represent only one aspect of environmental conditions.

Differential emphasis of developmental theories

1. Physical growth and development can be viewed differently. Such distinction was made at the Gessell Institute. Growth is because of maturation and implies an increase in size. Development is a focus on increase in complexity and diversity rather than just an increase in size.

2. Cognitive development theories emphasize how the intellectual processes develop. Such theorists are more interested in how individual processes develop rather than how individuals differ.
3. Moral development is generally an investigation of the bases for moral decision making
4. Social development is a focus on how people develop and relate to other members of society.
5. Personality theories are focused on traits, attitudes and behaviors and attempts to explain those by referring to past and present environmental influences, social conditions and genetic factors. Theorists tend to be interested in labels.
6. Learning theorists attempt to explain the factors that influence learning. Teaching theory is largely based on learning theory.
7. Behavioral theory is focused on measurable, observable behavior. Behavior is a product of the environment. Theorists tend not to label behavior.
8. Psychoanalytic theories are focused on the unconscious influence on human behavior and development.
9. Humanists focus on explanations dealing with experience and consciousness.

Boeree suggests that ideas related to human development can be organized into ten categories.

1. Free will versus determinism
2. Uniqueness versus universality
3. Physiological versus purposive motivation
4. Conscious versus unconscious motivation
5. Nature versus nurture
6. Stage versus non-stage

7. Cultural determination versus cultural transcendence
8. Early or late personality formation
9. Continuous versus discontinuous understanding of mental illness
10. Optimism versus pessimism

Boeree also suggested that human development theories could be classified into three types, which are:

1. Psychoanalytical
2. Behavioristic
3. Humanistic

Another way of organizing human developmental theories of the six categories that follow.

1. Psychobiological model-basic assumption is that human development is shaped by a combination of biological and neuropsychological processes
2. Social and behavioral model-focus on understanding social factors and modifying behavior on the assumption that there was inappropriate conditioning
3. Cognitive models-focus on links between mental factors, emotions, actions and stress individual aspects
4. Psychodynamic models-focus on unconscious, transference and object relations-instinctive urges and drives cause us to behave in certain ways
5. Humanistic models-focus on individuals' values, choices, autonomy, responsibility and potential for change on individuals' values, choices, autonomy, responsibility and potential for change
6. Systemic and group models-focus is on interaction of factors, family is important, problems are socially embedded, development in groups is important

Freud

1. Credited for developing psychoanalytic and psychosexual theories
2. Deterministic view, biological forces motivate personality
3. Study of mental processes
4. Based largely on the assumption that the conscious and unconscious are on conflict
5. Through free association, unconscious thoughts become conscious
6. Created idea of ego (conscious system), superego (conscience system) and the id (unconscious system)
7. Describes the oral, anal and phallic stages

Erikson

1. Viewed human development as a series of psychosocial stages across the life span
2. Resolving the crises between internal maturational forces and external social events is what advances people from one stage to another.
3. Erikson's eight stages
   a. Trust versus mistrust
   b. Autonomy versus shame
   c. Initiative versus guilt
   d. Industry versus inadequacy
   e. Identity versus role confusion
   f. Intimacy versus isolation
   g. Generatively versus stagnation
   h. Integrity versus despair.

4. Erikson believed in the principle of epigenesis, which is that all living things grow according to a specific maturational plan.
5. He also embraced the idea that psychosocial conflict-conflict between an individual's search for identity and the desire for social conformity is mediated by the ego. If the crisis is resolved, one can move to the next stage.

Piaget

1. A cognitive development theorists
2. He suggested that all human beings strive to maintain a sense of order or equilibrium and that everyone has an innate tendency to adapt to the environment. When the mental structure or knowledge is no longer sufficient to meet the demands of the environment, humans experience disequilibruim and then use cognitive adaptive measures to regain order.
3. Adaptation involves the processes of accommodation and assimilation. Accommodation is the reorganization of cognitive structures to compensate for changing environmental demands. Assimilation refers to the incorporation of new objects or information into existing cognitive structures.
4. Piaget stages are
   a. Sensorimotor-birth to 2 years
   b. Preoperational-2 to 6 years
   c. Concrete-operational-7 to 11 years
   d. Formal operations-12 + years
5. Piaget suggested that healthy cognitive development is the product of four factors
   a. Maturation-biological processes associated with physical and neurological growth

b. Experience-direct interaction with the environment

c. Transmission of attitudes and cultural values

d. Equilibration-an underlying self-regulating process

Gardner

Howard Gardner proposed in the mid-eighties that there was more than one type of intelligence. He suggested in his Frames of mind the following types of intelligence.

1. Linguistic-meanings of words important-good with grammar and semantics
2. Musical-special facility to understand and learn music
3. Logical-mathematical-understanding concepts and calculations
4. Spatial-good with objects in space-find way around easily-see physical relationships
5. Bodily kinesthetic-able to illustrate many emotions-able to manipulate objects where fine motor skills are needed-using ones body effectively
6. Intrapersonal-able to get inside self-explore depths of desires, emotions and purpose
7. Interpersonal-ability to observe others' moods, temperaments, motivations and intentions and relate effectively Goleman.

Daniel Goleman, building upon the work of Gardner, suggested that combining Gardner's intrapersonal and interpersonal would constitute emotional intelligence. Characteristics of those with emotional intelligence follow.

1. Recognize own emotions and effect on others
2. Assess self accurately
3. Self-confidence

4. Self-control
5. Adaptable and innovative
6. High achievement drive
7. Proactive and persistent
8. Service orientation
9. Effective communicators
10. As leaders, able to inspire and guide

Kohlberg

1. Development of moral reasoning
2. Moral judgments are the products of an individual's thinking about moral problems
3. Cognitive development is a prerequisite for moral development, but that must be combined with social experiences to stimulate the mental processes to move to the next stage.
4. Individuals progress through the stages in a fixed order and each stage is built on the previous stage
5. Behavior is not the distinguishing factor between stages-the reasoning underlying how a moral problem is processed determines movement
6. Kohlberg's stages
   a. Level I-preconventional morality-consists of the stages of obedience/punishment and individualism/exchange
   b. Level II-conventional morality-consist of good interpersonal relationships and maintaining social order
   c. Level III-postconventional morality-consists of social contract/individual rights and universal principles
7. Kohlberg developed a series of moral dilemmas to study moral reasoning

Bandura

1. A social learning theorists
2. Individuals are actively involved in the learning process
3. Focused on the process of change more than Piaget, Erikson and Kohlberg
4. Basic assumptions include
   a. A child can acquire new behaviors without direct experience-vicariously
   b. Individual is seen as operating on the environment
   c. Learning is a process of interaction among behavior, personal factors and the environment
   d. Acknowledges role of heredity, but feels the environment is more important
5. Methodology for testing ideas is based on controlled studies-manipulation of certain variables
6. Recognizes the role of emotion and motivation -rather than just logical processing in the learning process

Skinner

1. Clearly a behaviorist
2. Theories are based on the fundamental assumptions that human behavior is shaped by the nature of the consequences of that behavior
3. Quest for freedom is not self-directed, but an avoidance of aversive stimuli
4. Cultures that help people satisfy needs and avoid dangers will grow
5. Culture is the driving force behind human behavior
6. The self made person is a product of the culture in which the person lives

Maslow

1. Suggested that mankind was motivated to develop by a series of needs which are:
   a. Physiological
   b. Safety and Security
   c. Love and Belonging
   d. Esteem
   e. Self-Actualization

Loevinger

1. Developed a theory she called Ego Development. That is an increase in differentiation of the conception of self within the environment, involving a movement away from manipulative, self-protective behaviors to an integrated understanding of self and others.
2. Not a stage theory because people can be at different places at different ages
3. Her theory has ten levels with some being transitional stages. They are:
   a. Presocial Stage-ego formation begins-starts to separate from environment
   b. Symbiotic Stage-maintains dependent relationship-continues separation from environment
   c. Impulsive Stage-relations with others are dependent, but exploitative-controlled with immediate rewards or punishment-actions and people are judged in absolutes
   d. Self-Protective Stage-rules are recognized and used to one's advantage-manipulative and exploitative characterize persons who stay at this stage

e. Conformist Stage-behavior based on expectations of others and socially approved norms-belonging is important

f. Self-Aware-a transitional stage-desire for others to approve, appreciate and respect-growing self-confidence-self-awareness-self-evaluated standards

g. Conscientious Stage-relationships take on new depth-rules internalized-guilt comes from breaking rules-achievement measured by own standards in important

h. Individualistic Stage-a transitional stage-show concern for others personal growth and respect for others freedom

i. Autonomous Stage-growing awareness of how to respect and enhance development of others-interpersonal relationships are cherished -a sense of individuality-knowledge of emotional dependence on others-increased tolerance for contradictory relationships

j. Integrated Stage-have become self-actualized in Maslow's terms-no more that about 1% reach this stage-wants individuality and creative integration of his interpersonal life and the values placed upon life

Harvey, Hunt and Schroeder

1. Developed conceptual levels of development which are:

   a. Unilateral Dependence-difficulty understanding dissimilar concepts-difficulty with ambiguity-sees things in terms of absolutes

   b. Negative Independence-opposes control by external rules or conditions-questions external opposition-tests limits

   c. Conditioning Dependence and Mutuality-begins to establish mutual relationship with environment-can combine and compare different conditions-potential for self-reflection

d. Interdependence-can order, combine, and evaluate different concepts in different ways-capable of interdependent relations with environment and others in that environment-adapts to complex and changing environment What are some ideas you have for the content of the depth and application sections?

DEVELOPMENT OF CONGNITIVE THEORY

Generally it is referred to the changes which occur to a person's cognitive structures, abilities, and processes. The most widely known theory of childhood cognitive development was proposed by Jean Piaget in 1969. He proposed the idea that cognitive development consisted of the development of logical competence, and that the development of this competence consists of four major stages:

— sensori-motor

— preoperational

— concrete operational

— formal operational

He also argued that a child's cognitive performance depended more on the stage of development he was in than on the specific task being performed. More recent studies have cast some doubt on Piaget's theory of homogeneous performance within a given stage. Instead, it is now believed that performance varies greatly within each stage and depends more on the acquisition and development of language, perception, decision rules, and real-world knowledge for any individual child.

Cognitive styles

Cognitive styles refer to the preferred way an individual processes information. Unlike individual differences in abilities (e.g., Gardner, Guilford, Sternberg) which describe peak performance, styles describe a person's typical mode of thinking, remembering or problem solving. Furthermore, styles are usually considered to be bipolar dimensions whereas abilities are unipolar (ranging from zero to a maximum value). Having more of an ability is usually considered beneficial while having a particular cognitive style simply denotes a tendency to behave in a certain manner.

Cognitive style is a usually described as a personality dimension which influences attitudes, values, and social interaction. A number of cognitive styles have been identified and studied over the years. Field independence versus field dependence is probably the most well known style. It refers to a tendency to approach the environment in an analytical, as opposed to global, fashion. At a perceptual level, field independent personalities are able to distinguish figures as discrete from their backgrounds compared to field dependent individuals who experience events in an undifferentiated way.

In addition, field dependent individuals have a greater social orientation relative to field independent personalities. Studies have identified a number connections between this cognitive style and learning. For example, field independent individuals are likely to learn more effectively under conditions of instrinstic motivation (e.g., self-study) and are influenced less by social reinforcement. Other cognitive styles that have been identified include:

— *scanning* - differences in the extent and intensity of attention resulting in variations in the vividness of experience and the span of awareness

— *leveling versus sharpening* - individual variations in remembering that pertain to the distinctiveness of memories and the tendency to merge similar events

— *reflection versus impulsivity* - individual consistencies in the speed and adequacy with which alternative hypotheses are formed and responses made

— *conceptual differentiation* - differences in the tendency to categorize perceived similarities among stimuli in terms of separate concepts or dimensions

Learning styles specifically deal with characteristic styles of learning. Kolb proposes a theory of experiential learning that involves four principal stages: concrete experiences (CE), reflective observation (RO), abstract conceptualization (AC), and active experimentation (AE). The CE/AC and AE/RO dimensions are polar opposities as far as learning styles are concerned and Kolb postulates four types of learners (divergers, assimilators, convergers, and accommodators) depending upon their position on these two dimensions. For example, an accommodater prefers concrete experiences and active experimentation (AE, CE).

Pask has described a learning style called serialist versus holist. Serialists prefer to learn in a sequential fashion, whereas holists prefer to learn in a hierarchial manner (i.e., top-down). Theoretically, cognitive and learning styles could be used to predict what kind of instructional strategies or methods would be most effective for a given individual and learning task.

Metacognition

Metacognition is the process of thinking about thinking. Flavell describes it as follows: "Metacognition refers to one's knowledge concerning one's own cognitive processes or anything related to them, e.g., the learning-relevant properties of information or data. For example, I am engaging in metacognition if I notice that I am having more trouble learning A than B; if it strikes me that I should double check C before accepting it as fact.".

Flavell argued that metacognition explains why children of different ages deal with learning tasks in different ways, i.e., they have developed new strategies for thinking. Research studies seem to confirm this conclusion; as children get older they demonstrate more awareness of their thinking processes.

Metacognition has to do with the active monitoring and regulation of cognitive processes. It represents the "executive control" system that many cognitive theorists have included in their theories (e.g., Miller, Newell & Simon, Schoenfeld). Metacognitive processes are central to planning, problem-solving, evaluation and many aspects of language learning.

Metacognition is relevant to work on cognitive styles and learning strategies in so far as the individual has some awareness of their thinking or learning processes. The work of Piaget is also relevant to research on metacognition since it deals with the development of cognition in children.

The Cognitive System

Cognition can be defined as "the act or process of knowing in the broadest sense; specifically, an intellectual process by which knowledge is gained from perception or ideas". Cognition is central to the development of psychology as a scientific discipline. The establishment of Wilhelm Wundt's laboratory in 1879 to study human thought processes is often used as the beginning of modern psychology. Cognitive psychology is one of the major approaches within psychology and can be contrasted with the behavioral view (a focus on observable behavior), a psychoanalytic view (a focus on the unconscious), a humanistic view (a focus on personal growth and interpersonal relationships) and a social cognitive view (a focus on the social environment as it impacts personal qualities such as thinking and feeling.)

An important distinction between the behavioral and cognitive or humanistic views is the importance of feedback. For behaviorists, the most important feedback comes in the form of the application of consequences from the environment. The cognitive, as well humanistic, would focus on the importance of internal feedback. The social cognitive view considers both types. There are a variety of perspectives and emphases within cognitive psychology that are currently impacting educators' thinking about how to improve the teaching/learning process. The Information Processing approach focuses on the study of the structure and function of mental processing within specific contexts, environments, or ecologies .

Benjamin Bloom and his colleagues developed the Taxonomy of the Cognitive Domain as a way to classify the variety of educational objectives related to what and how we know. Researchers in the area of intelligence study how human beings learn from experience, reason well, remember important information, and adapt to the environment. Jean Piaget's theory of cognitive development describes the process and stages by which human beings develop the capacity to engage in abstract symbolic thought, one of the distinguishing features of human activity. Piaget's theory is often contrasted with the views of Jerome Bruner and Lev Vygotsky.

Several different areas of inquiry provide an opportunity to test out these different theories. For example, in the area of critical thinking researchers study how we apply our cognitive processes to evaluating arguments (propositions) and making decisions. On the other hand, in the area of creative thinking researchers study how we generate ideas and alternatives that don't fit the "norm." These two areas are often contrasted as the difference between convergent thinking (thinking pattern that we use when we want to narrow down and evaluate ideas) and divergent thinking (thinking pattern we use when we want to expand or develop new ideas). A similar comparison is between left-brain and right-brain orientations in the literature on brain lateralization dominance.

Metacognition is another area in cognition that draws from a number of different perspectives and is the study of how we develop knowledge about one's own cognitive system. The area of study methods, such as SQ4R, provides information about how we can be most effective and efficient during the process of learning.

COGNETIVE APPROACH

"...focuses on how we process, store, and use information and how this information influences what we attend to, perceive, learn, remember, believe, and feel"

Piaget's Theory of Cognitive Development

Jean Piaget (1896-1980) was one of the most influential researchers in the area of developmental psychology during the 20th century. Piaget originally trained in the areas of biology and philosophy and considered himself a "genetic epistemologist." He was mainly interested in the biological influences on "how we come to know." He believed that what distinguishes human beings from other animals is our ability to do "abstract symbolic reasoning."

Piaget's views are often compared with those of Lev Vygotsky (1896-1934), who looked more to social interaction as the primary source of cognition and behavior. This is somewhat similar to the distinctions made between Freud and Erikson in terms of the development of personality. The writings of Piaget and Vygotsky, along with the work of John Dewey, Jerome Bruner and Ulrick Neisser form the basis of the constructivist theory of learning and instruction.

Piaget became interested in how children think. He noticed that young children's answers were qualitatively different than older children which suggested to him that the younger ones were not dumber (a quantitative position since as they got older and had more experiences they would get smarter) but, instead, answered the questions differently than their older peers because they thought differently. There are two major aspects to his theory: the process of coming to know and the stages we move through as we gradually acquire this ability.

Process of Cognitive Development

As a biologist, Piaget was interested in how an organism adapts to its environment (Piaget described as intelligence.) Behavior (adaptation to the environment) is controlled through mental organizations called schemes that the individual uses to represent the world and designate action. This adaptation is driven by a biological drive to obtain balance between schemes and the environment (equilibration).

Piaget hypothesized that infants are born with schemes operating at birth that he called "reflexes." In other animals, these reflexes control behavior throughout life. However, in human beings as the infant uses these reflexes to adapt to the environment, these reflexes are quickly replaced with constructed schemes.

Piaget described two processes used by the individual in its attempt to adapt: assimilation and accomodation. Both of these processes are used thoughout life as the person increasingly adapts to the environment in a more complex manner.

Assimilation is the process of using or transforming the environment so that it can be placed in preexisting cognitive structures. Accomodation is the process of changing cognitive structures in order to accept something from the environment. Both processes are used simultaneously and alternately throughout life. An example of assimilation would be when an infant uses a sucking schema that was developed by sucking on a small bottle when attempting to suck on a larger bottle. An example of accomodation would be when the child needs to modify a sucking schema developed by sucking on a pacifier to one that would be successful for sucking on a bottle.

As schemes become increasingly more complex (i.e., responsible for more complex behaviors) they are termed structures. As one's structures become more complex, they are organized in a hierarchical manner (i.e., from general to specific).

Stages of Cognitive Development

Piaget identified four stages in cognitive development:

Sensorimotor stage (Infancy)

In this period (which has 6 stages), intelligence is demonstrated through motor activity without the use of symbols. Knowledge of the world is limited (but developing) because its based on physical interactions / experiences. Children acquire object permanence at about 7 months of age (memory). Physical development (mobility) allows the child to begin developing new intellectual abilities. Some symbollic (language) abilities are developed at the end of this stage.

Pre-operational stage

In this period (which has two substages), intelligence is demonstrated through the use of symbols, language use matures, and memory and imagination are developed, but thinking is done in a nonlogical, nonreversable manner. Egocentric thinking predominates

Concrete operational stage

In this stage (characterized by 7 types of conservation: number, length, liquid, mass, weight, area, volume), intelligence is demonstarted through logical and systematic manipulation of symbols related to concrete objects. Operational thinking develops (mental actions that are reversible). Egocentric thought diminishes.

Formal operational stage

In this stage, intelligence is demonstrated through the logical use of symbols related to abstract concepts. Early in the period there is a return to egocentric thought. Only 35% of high school graduates in industrialized countries obtain formal operations; many people do not think formally during adulthood.

Many pre-school and primary programs are modeled on Piaget's theory, which, as stated previously, provides part of the foundation for constructivist learning. Discovery learning and supporting the developing interests of the child are two primary instructional techniques. It is recommended that parents and teachers challenge the child's abilities, but NOT present material or information that is too far beyond the child's level. It is also recommended that teachers use a wide variety of concrete experiences to help the child learn (e.g., use of manipulatives, working in groups to get experience seeing from another's perspective, field trips, etc).

Piaget's research methods were based primarily on case studies [they were descriptive]. While some of his ideas have been supported through more correlational and experimental methodologies, others have not. For example, Piaget believed that biological development drives the movement from one cognitive stage to the next. Data from cross-sectional studies of children in a variety of western cultures seem to support this assertion for the stages of sensorimotor, preoperational, and concrete operations.

However, data from similar cross-sectional studies of adolescents do not support the assertion that all individuals will automatically move to the next cognitive stage as they biologically mature. Data from adult populations provides essentially the same result: Between 30 to 35% of adults attain the cognitive development stage of formal operations. For formal operations, it appears that maturation establishes the basis, but a special environment is required for most adolescents and adults to attain this stage.

Piaget's Stage Theory of Development

Piaget was among other things, a psychologist who was interested in cognitive development. After observation of many children, he posited that children progress through 4 stages and that they all do so in the same order. These four stages are described below.

The Sensorimotor Period (birth to 2 years)

During this time, Piaget said that a child's cognitive system is limited to motor reflexes at birth, but the child builds on these reflexes to develop more sophisicated procedures. They learn to generalize their activities to a wider range of situations and coordinate them into increasingly lengthy chains of behaviour.

PreOperational Thought (2 to 6/7 years)

At this age, according to Piaget, children acquire representational skills in the areas mental imagery, and especially language. They are very self-oriented, and have an egocentric view; that is, preoperational chldren can use these representational skills only to view the world from their own perspective.

Concrete Operations (6/7 to 11/12 years)

As opposed to Preoperational children, children in the concrete operations stage are able to take another's point of view and take into account more than one perspective simultaneously. They can also represent transformations as well as static situations. Although they can understand concrete problems, Piaget would argue that they cannot yet perform on abstract problems, and that they do not consider all of the logically possible outcomes.

Formal Operations (11/12 to adult)

Children who attain the formal operation stage are capable of thinking logically and abstractly. They can also reason theoretically. Piaget considered this the ultimate stage of development, and stated that although the children would still have to revise their knowledge base, their way of thinking was as powerful as it would get.

It is now thought that not every child reaches the formal operation stage. Developmental psychologists also debate whether children do go through the stages in the way that Piaget postulated. Whether Piaget was correct or not, however, it is safe to say that this theory of cognitive development has had a tremendous influence on all modern developmental psychologists.

VYGOTSKY'S APPROACH TO COGNITIVE DEVELOPMENT

One of the major theoretical advances of Vygotsky's approach to cognitive development was his thesis that human mental functions were social in origins. In making this claim, Vygotsky was confronted with the difficulty to reconciliate it with the existing fact that newborn infants already possess certain mental functions. Vygotsky's answer to the problem was the introduction of an important distinction between lower mental functions (LMF's) and higher mental functions (HMF's).

The relationships between the two in Vygotsky's theory was not strictly determined. In some cases LMF can be a prerequisite for the development of an appropriate HMF (i.e., unmediated memory can be developed in mediated and voluntarily controlled memory), in other cases HMF's exist in the intersubjective form and are merely learned by the child in the process of education and shared activities (i.e., writing or reading skills). In both cases Vygotsky applied Hegelian developmental scheme to the development of cognitive skills, according to which any cognitive function goes through three major stages, in which it exists at first 'in itself', then 'for others' and finally 'for itself.'

For example, Vygotsky portrayed the development of an indicatory gesture in infancy as a series of stages. In the beginning it is just an unsuccessful grasping movement directed towards a desired object. As such this is not yet an indication, but it can acquire the meaning if interpreted appropriately by the child's caregivers. At this stage the grasping movement becomes mediated by the social environment and acquires a social meaning 'help me to get this' which is quickly absorbed by the child who begins to use it both for the purpose of communication with the caregivers and for achieving his or her practical goals.

While doing this, the child can still be unaware of the fact that he or she is exploiting the gesture as a social signal. Still later (my interpretation of Vygotsky's text) this 'gesture-forothers' can become a kind of a 'tool' by which the child would exercise control over his or her own actions and behaviour, for instance, in order to pinpoint a certain fragment of a picture and concentrate his or her attention on it. This time the child is fully aware that what he or she is doing with his or her forefinger (or whatever may substitute for it) is a special act designed not to let his or her attention to wander around the picture but to stick to a certain elected point. This is the stage when the indicatory gesture exists 'for itself' or, strictly speaking, for the child who utilises it being at the same time fully aware of that.

On a more broad scale, the development of human mental functions is viewed by Vygotsky as their transition from their original lower mental functions form into higher mental functions form, with differences between the two being drawn along four major criteria: origins, structure, the way of functioning and the relation to other mental functions. By origins, most lower mental functions are genetically inherited, by structure they are unmediated, by functioning they are involuntary, and with regard to their relation to other mental functions they are isolated individual mental units.

In contrast, a higher mental function is socially acquired, mediated by social meanings, voluntarily controlled and exists as a link in a broad system of functions rather than as an individual unit. However, even in Vygotsky's time there existed some experimental data that posed difficulties for this kind of developmental approach. One of the most extraordinary claims was made by Gestalt-psychologists; according to this claim, some universal structural laws are innate for human perception. These laws (like the 'law of common fate') are not acquired through learning; rather, they are present in the infant from birth and do not change with age.

In particular, Folkelt suggested that infant's perception from birth had structural and 'orthoscopic' character (the implication of this claim was, for instance, that infants' possess the inherent capacity to the constancy of perception). No wonder that Vygotsky (1982) was strongly opposed to this view. His major objection was theoretically, rather then empirically, based: if infants have an inherent capacity to the constancy of perception, then where are we to find the development?

In other words, if the final stage of perceptual development is present from the outset, the concept of development becomes superfluous. Searching for evidence to back his claim Vygotsky addressed Helmgoltz's early memories from his childhood in which he suggested that orthoscopic perception was not inborn but had to develop through experience. Although Vygotsky himself qualified the Helmgoltz's report as a shaky evidence he, nevertheless, accepted it as one of the proofs in favour of the assumption of the acquired nature of orthoscopic perception.

Yet recent developments in infancy studies have demonstrated the striking perfection of infant's perception. Let us mention but a few of them: Bower (1979) reported the data suggesting that 3-week-old infants reveal certain 'understanding' of the 'law of common fate'; Slater, Morrison & Rose found newborns to be able to distinguish between main archetypical figures (like a cross and a circle); Gibson & Walker demonstrated that 1-month-old infants were able to perceive consistence (rigid vs. elastic) and to transfer the information of object consistence from tactile into visual modality; again, Bower and later Slater & Morrison found that 8- weeks-old and newborn infants could perceive objects as constant in shape; Baillargeon established that 3 1/2 - and 4 1/2-month- old could 'understand' some physical properties of a solid body such as its impermeability for another solid body.

So, was Vygotsky wrong in his denial that young infants could possibly possess such complex psychological abilities as, for instance, the capacity to perceive an object as constant in shape or size? The affirmative answer, which seems inevitable, can not, however, be given but with serious reservations.

.Firstly, the way the infants' early capacities are presented and discussed by many authors provokes questions. A characteristic feature of most of recent accounts on the problem is that the infants' early cognitive skills are portrayed in exactly the same terms as are similar capacities in adults: for instance, the infants are supposed to be able to 'infer' that a physical object without a support would fall down rather than hang in the air, they can 'understand' that a solid object can not go through another solid object, they are able to 'appreciate' object permanence or object constancy, and so on. It is not that the qualitative difference between psychological functions of infants and those of adults is openly denied; rather, it is taken for granted that either these differences do not extend to cover the capacities in question or they are not really important.

In a result, the question about what exactly distinguishes, for instance, the 5-months-old infant's behaviour testifying that the infant can understand object permanence from a similar behaviour of an adult person is very rarely asked, and when it is asked the usual answer is that the difference is nothing but a scope of applicability of the cognitive skill: if an infant can apply the permanence rule to a limited number of cases, an adult person is able to generalize the rule to a much larger number of observable physical events.

In other words, a careful reading discovers that the development of cognitive skills is indeed interpreted by many as a quantitative perfection of the early acquired (or genetically transmitted) capacity rather than a series of qualitative changes that the capacity has to go through in order to reach its higher stage. Therefore, despite the fact that the Vygotsky's answer may have been wrong, his question was correct: indeed, where is (and what is) cognitive development if major psychological capacities in their almost completed form are here in the first few months of life?

Secondly, if we look at the potential content of the Vygotsky's answer, rather then at it's literally meaning, we can see that it was rather contradictive. On the one hand, Vygotsky denied the inherent character of the constancy of size on the ground that it was an internally complex psychological quality and hence it must be a socially formed quality. On the other hand, if we look at the criteria that distinguish LMF's from HMF's, we won't find the internal complexity among them.

Indeed, as it was already noted, in contrast to LMF's which are inherent, involuntary, unmediated and isolated one from another, HMF's are socially created, voluntarily controlled, semiotically mediated and united in systems with other functions. Clearly, there is no claim here from which it would follow that LMF has to lack the internal complexity and perfection which is normally attributed to adults, but not to newborns and young infants.

Taking this into consideration and ignoring Vygotsky's misleading, although sincere, disbelief in the possibility of inherent and genetically transmitted complex mental function, we can assume that Vygotsky's distinction between LMF's and HMF's still has something to offer to the recent findings in infancy studies. Clearly, the extraordinary capacities of infants that are now being displayed in a growing number of studies, although complex, are still LMF's and have to go through the route of development (becoming semiotically mediated, voluntarily controlled, and united in systems with other mental functions) that so ingeniously was outlined by Vygotsky.

Congnitive Development Views of Vygotsky, Piaget, and Bruner

Lev Vygotsky, Jean Piaget, and Jerome Bruner have emphasized cognitive development as being intimately linked to the brain's construction of knowledge within a social context. Their work has been instrumental in providing a foundation for multiage grouping

Lev Vygotsky

Vygotsky asserts that the most fruitful experience in a child's education is his or her collaboration with more skilled partners. Vygotsky explains that the more experienced partner provides help in the way of an intellectual scaffold, which allows the less experienced learner to accomplish more complex tasks than may be possible alone. Within a multiage setting, many opportunities exist for interaction between children of different ages, experiences, and developmental levels.

Jean Piaget

Piaget considers the most critical factor in a child's cognitive development to be interaction with peers. Interaction lends opportunities for the child to have cognitive conflict, which results in arguing or debating with peers. This type of interaction requires children to decenter, or consider another person's point of view. Piaget observes that children are most challenged in their thinking when they are with peers, because they all are on an equal footing and are more free to confront ideas than when interacting with adults. However, when children are too similar in their thinking, there may be little to debate about, resulting in fewer developmental gains. The multiage setting maximizes a child's opportunity for cognitive conflict because it brings together children at a variety of developmental levels. For further information, refer to Thinking About Piaget in Relationship to the Mixed-Age Classroom.

Jerome Bruner

Bruner observes that the process of constructing knowledge of the world is not done in isolation but rather within a social context. The child is a social being and, through social life, acquires a framework for interpreting experiences. Bruner also notes that "there is no unique sequence for all learners, and the optimum in any particular case will depend upon a variety of factors, including past learning, stage of development, nature of the material, and individual differences".

Effective curriculum then, must provide many opportunities and choices of children.. Within the multiage setting, opportunities exist for children to make choices about their learning experiences. In addition, the variety of teaching methods used in the multiage classroom provides opportunities for children to construct knowledge in a multitude of ways.

Kohilberg's Stages of Moral Development

An outstanding example of research in the Piagetian tradition is the work of Lawrence Kohlberg. Kohlberg has focused on moral development and has proposed a stage theory of moral thinking which goes well beyond Piaget's initial formulations. Kohlberg, who was born in 1927, grew up in Bronxville, New York, and attended the Andover Academy in Massachusetts, a private high school for bright and usually wealthy students. He did not go immediately to college, but instead went to help the Israeli cause, in which he was made the Second Engineer on an old freighter carrying refugees from parts of Europe to Israel.

After this, in 1948, he enrolled at the University of Chicago, where he scored so high on admission tests that he had to take only a few courses to earn his bachelor's degree. This he did in one year. He stayed on at Chicago for graduate work in psychology, at first thinking he would become a clinical psychologist. However, he soon became interested in Piaget and began interviewing children and adolescents on moral issues. The result was his doctoral dissertation (1958a), the first rendition of his new stage theory.

Kohlberg is an informal, unassuming man who also is a true scholar; he has thought long and deeply about a wide range of issues in both psychology and philosophy and has done much to help others appreciate the wisdom of many of the "old psychologists," such as Rousseau, John Dewey, and

James Mark Baldwin. Kohlberg has taught at the University of Chicago and, since 1968, has been at Harvard University.

Piaget's Stages of Moral Judgement

Piaget studied many aspects of moral judgment, but most of his findings fit into a two-stage theory. Children younger than 10 or 11 years think about moral dilemmas one way; older children consider them differently. As we have seen, younger children regard rules as fixed and absolute. They believe that rules are handed down by adults or by God and that one cannot change them. The older child's view is more relativistic. He or she understands that it is permissible to change rules if everyone agrees. Rules are not sacred and absolute but are devices which humans use to get along cooperatively.

At approximately the same time—10 or 11 years—children's moral thinking undergoes other shifts. In particular, younger children base their moral judgments more on consequences, whereas older children base their judgments on intentions. When, for example, the young child hears about one boy who broke 15 cups trying to help his mother and another boy who broke only one cup trying to steal cookies, the young child thinks that the first boy did worse. The child primarily considers the amount of damage—the consequences—whereas the older child is more likely to judge wrongness in terms of the motives underlying the act.

There are many more details to Piaget's work on moral judgment, but he essentially found a series of changes that occur between the ages of 10 and 12, just when the child begins to enter the general stage of formal operations.

Intellectual development, however, does not stop at this point. This is just the beginning of formal operations, which continue to develop at least until age 16. Accordingly, one might expect thinking about moral issues to continue to develop throughout adolescence. Kohlberg therefore interviewed both children and adolescents about moral dilemmas, and he did find stages that go well beyond Piaget's. He uncovered six stages, only the first three of which share many features with Piaget's stages.

Kohlberg had classified the various responses into stages, he wanted to know whether his classification was *reliable*. In particular, he. wanted to know if others would score the protocols in the same way. Other judges independently scored a sample of responses, and he calculated the degree to which all raters agreed. This procedure is called *interrater reliability*. Kohlberg found these agreements to be high, as he has in his subsequent work, but whenever investigators use Kohlberg's interview, they also should check for interrater reliability before scoring the entire sample.

Six Stages sof Kohlberg

Level 1. Preconventional Morality

Stage 1. Obedience and Punishment Orientation. Kohlberg's stage 1 is similar to Piaget's first stage of moral thought. The child assumes that powerful authorities hand down a fixed set of rules which he or she must unquestioningly obey. To the Heinz dilemma, the child typically says that Heinz was wrong to steal the drug because "It's against the law," or "It's bad to steal," as if this were all there were to it. When asked to elaborate, the child usually responds in terms of the consequences involved, explaining that stealing is bad "because you'll get punished".

Although the vast majority of children at stage 1 oppose Heinz's theft, it is still possible for a child to support the action and still employ stage 1 reasoning. For example, a child might say, "Heinz can steal it because he asked first and it's not like he stole something big; he won't get punished". Even though the child agrees with Heinz's action, the reasoning is still stage 1; the concern is with what authorities permit and punish.

Kohlberg calls stage 1 thinking "preconventional" because children do not yet speak as members of society. Instead, they see morality as something external to themselves, as that which the big people say they must do.

Stage 2. Individualism and Exchange. At this stage children recognize that there is not just one right view that is handed down by the authorities. Different individuals have different viewpoints. "Heinz," they might point out, "might think it's right to take the drug, the druggist would not." Since everything is *relative,* each person is free to pursue his or her *individual* interests. One boy said that Heinz might steal the drug if he wanted his wife to live, but that he doesn't have to if he wants to marry someone younger and better-looking. Another boy said Heinz might steal it because maybe they had children and he might need someone at home to look after them. But maybe he shouldn't steal it because they might put him in prison for more years than he could stand.

What is right for Heinz, then, is what meets his own self-interests. You might have noticed that children at both stages 1 and 2 talk about punishment. However, they perceive it differently. At stage 1 punishment is tied up in the child's mind with wrongness; punishment "proves" that disobedience is wrong. At stage 2, in contrast, punishment is simply a risk that one naturally wants to avoid.

Although stage 2 respondents sometimes sound amoral, they do have some sense of right action. This is a notion of *fair exchange* or fair deals. The philosophy is one of returning favors—"If you scratch my back, I'll scratch yours." To the Heinz story, subjects often say that Heinz was right to steal the drug because the druggist was unwilling to make a fair deal; he was "trying to rip Heinz off," Or they might say that he should steal for his wife "because she might return the favor some day".

Respondents at stage 2 are still said to reason at the preconventional level because they speak as isolated individuals rather than as members of society. They see individuals exchanging favors, but there is still no identification with the values of the family or community.

Level II. Conventional Morality

Stage 3. Good Interpersonal Relationships. At this stage children—who are by now usually entering their teens—see morality as more than simple deals. They believe that people should live up to the expectations of the family and community and behave in "good" ways. Good behavior means having good motives and interpersonal feelings such as love, empathy, trust, and concern for others. Heinz, they typically argue, was right to steal the drug because "He was a good man for wanting to save her," and "His intentions were good, that of saving the life of someone he loves." Even if Heinz doesn't love his wife, these subjects often say, he should steal the drug because "I don't think any husband should sit back and watch his wife die".

If Heinz's motives were good, the druggist's were bad. The druggist, stage 3 subjects emphasize, was "selfish," "greedy," and "only interested in himself, not another life." Sometimes the respondents become so angry with the druggist that they say that he ought to be put in jail. A typical stage 3 response is that of Don, age 13:

Don defines the issue in terms of the actors' character traits and motives. He talks about the loving husband, the unfair druggist, and the understanding judge. His answer deserves the label "conventional "morality" because it assumes that the attitude expressed would be shared by the entire community—"anyone" would be right to do what Heinz did.

As mentioned earlier, there are similarities between Kohlberg's first three stages and Piaget's two stages. In both sequences there is a shift from unquestioning obedience to a relativistic outlook and to a concern for good motives. For

Kohlberg, however, these shifts occur in three stages rather than two.

Stage 4. Maintaining the Social Order. Stage 3 reasoning works best in two-person relationships with family members or close friends, where one can make a real effort to get to know the other's feelings and needs and try to help. At stage 4, in contrast, the respondent becomes more broadly concerned with *society as a whole.* Now the emphasis is on obeying laws, respecting authority, and performing one's duties so that the social order is maintained. In response to the Heinz story, many subjects say they understand that Heinz's motives were good, but they cannot condone the theft. What would happen if we all started breaking the laws whenever we felt we had a good reason? The result would be chaos; society couldn't function. As one subject explained,

I don't want to sound like Spiro Agnew, law and order and wave the flag, but if everybody did as he wanted to do, set up his own beliefs as to right and wrong, then I think you would have chaos. The only thing I think we have in civilization nowadays is some sort of legal structure which people are sort of bound to follow. [Society needs] a centralizing framework. (Gibbs et al.). Because stage 4, subjects make moral decisions from the perspective of society as a whole, they think from a full-fledged member-of-society perspective.

You will recall that stage 1 children also generally oppose stealing because it breaks the law. Superficially, stage 1 and stage 4 subjects are giving the same response, so we see here why Kohlberg insists that we must probe into the reasoning behind the overt response. Stage 1 children say, "It's wrong to steal" and "It's against the law," but they cannot elaborate any further, except to say that stealing can get a person jailed. Stage 4 respondents, in contrast, have a conception of the function of laws for society as a whole—a conception which far exceeds the grasp of the younger child.

Level III. Postconventional Morality

Stage 5. Social Contract and Individual Rights. At stage 4, people want to keep society functioning. However, a smoothly functioning society is not necessarily a good one. A totalitarian society might be well-organized, but it is hardly the moral ideal. At stage 5, people begin to ask, "What makes for a good society?" They begin to think about society in a very theoretical way, stepping back from their own society and considering the rights and values that a society ought to uphold. They then evaluate existing societies in terms of these prior considerations. They are said to take a "prior-to-society" perspective.

Stage 5 respondents basically believe that a good society is best conceived as a social contract into which people freely enter to work toward the benefit of all They recognize that different social groups within a society will have different values, but they believe that all rational people would agree on two points. First they would all want certain basic *rights,* such as liberty and life, to be protected Second, they would want some *democratic* procedures for changing unfair law and for improving society.

In response to the Heinz dilemma, stage 5 respondents make it clear that they do not generally favor breaking laws; laws are social contracts that we agree to uphold until we can change them by democratic means. Nevertheless, the wife's right to live is a moral right that must be protected. Thus, stage 5 respondent sometimes defend Heinz's theft in strong language:

It is the husband's duty to save his wife. The fact that her life is in danger transcends every other standard you might use to judge his action. Life is more important than property. This young man went on to say that "from a moral standpoint" Heinz should save the life of even a stranger, since to be consistent, the value of a life means any life. When asked if the judge should punish Heinz, he replied:

Usually the moral and legal standpoints coincide. Here they conflict. The judge should weight the moral standpoint more heavily but preserve the legal law in punishing Heinz lightly.

Stage 5 subjects,- then, talk about "morality" and "rights" that take some priority over particular laws. Kohlberg insists, however, that we do not judge people to be at stage 5 merely from their verbal labels. We need to look at their social perspective and mode of reasoning. At stage 4, too, subjects frequently talk about the "right to life," but for them this right is legitimized by the authority of their social or religious group (e.g., by the Bible). Presumably, if their group valued property over life, they would too. At stage 5, in contrast, people are making more of an independent effort to think out what any society ought to value. They often reason, for example, that property has little meaning without life. They are trying to determine logically what a society ought to be like.

Stage 6: Universal Principles. Stage 5 respondents are working toward a conception of the good society. They suggest that we need to (a) protect certain individual rights and (b) settle disputes through democratic processes. However, democratic processes alone do not always result in outcomes that we intuitively sense are just. A majority, for example, may vote for a law that hinders a minority. Thus, Kohlberg believes that there must be a higher stage—stage 6—which defines the principles by which we achieve justice.

Kohlberg's conception of justice follows that of the philosophers Kant and Rawls, as well as great moral leaders such as Gandhi and Martin Luther King. According to these people, the principles of justice require us to treat the claims of all parties in an impartial manner, respecting the basic dignity, of all people as individuals. The principles of justice are therefore universal; they apply to all. Thus, for example, we would not vote for a law that aids some people but hurts others. The principles of justice guide us toward decisions based on an equal respect for all.

In actual practice, Kohlberg says, we can reach just decisions by looking at a situation through one another's eyes.

In the Heinz dilemma, this would mean that all parties—the druggist, Heinz, and his wife—take the roles of the others. To do this in an impartial manner, people can assume a "veil of ignorance", acting as if they do not know which role they will eventually occupy. If the druggist did this, even he would recognize that life must take priority over property; for he wouldn't want to risk finding himself in the wife's shoes with property valued over life. Thus, they would all agree that the wife must be saved—this would be the fair solution. Such a solution, we must note, requires not only impartiality, but the principle that everyone is given full and equal respect. If the wife were considered of less value than the others, a just solution could not be reached.

Until recently, Kohlberg had been scoring some of his subjects at stage 6, but he has temporarily stopped doing so, For one thing, he and other researchers had not been finding subjects who consistently reasoned at this stage. Also, Kohlberg has concluded that his interview dilemmas are not useful for distinguishing between stage 5 and stage 6 thinking. He believes that stage 6 has a clearer and broader conception of universal principles (which include justice as well as individual rights), but feels that his interview fails to draw out this broader understanding. Consequently, he has temporarily dropped stage 6 from his scoring manual, calling it a "theoretical stage" and scoring all postconventional responses as stage 5.

Theoretically, one issue that distinguishes stage 5 from stage 6 is civil disobedience. Stage 5 would be more hesitant to endorse civil disobedience because of its commitment to the social contract and to changing laws through democratic agreements. Only when an individual right is clearly at stake does violating the law seem justified. At stage 6, in contrast, a commitment to justice makes the rationale for civil disobedience stronger and broader.

Martin Luther King, for example, argued that laws are only valid insofar as they are grounded in justice, and that a commitment to justice carries with it an obligation to disobey unjust laws. King also recognized, of course, the general need for laws and democratic processes (stages 4 and 5), and he was therefore willing to accept the penalities for his actions. Nevertheless, he believed that the higher principle of justice required civil disobedience.

THEORITICAL ISSUES

Kohlberg, is a close follower of Piaget. Accordingly, Kohlberg's theoretical positions, including that on developmental change, reflect those of his mentor. Kohlberg says that his stages are not the product of maturation. That is, the stage structures and sequences do not simply unfold according to a genetic blueprint.

Neither, Kohlberg maintains, are his stages the product of socialization. That is, socializing agents (e.g., parents and teachers) do not directly teach new forms of thinking. Indeed, it is difficult to imagine them systematically teaching each new stage structure in its particular place in the sequence. The stages emerge, instead, from our own thinking about moral problems. Social experiences do promote development, but they do so by stimulating our mental processes. As we get into discussions and debates with others, we find our views questioned and challenged and are therefore motivated to come up with new, more comprehensive positions. New stages reflect these broader viewpoints.

We might imagine, for example, a young man and woman discussing a new law. The man says that everyone should obey it, like it or not, because laws are vital to social organization (stage 4). The woman notes, however, that some well-organized societies, such as Nazi Germany, were not particularly moral. The man therefore sees that some evidence contradicts his view. He experiences some cognitive conflict and is motivated to think about the matter more fully, perhaps moving a bit toward stage 5.

Kohlberg also sometimes speaks of change occurring through role-taking opportunities, opportunities to consider others' viewpoints. As children interact with others, they learn how viewpoints differ and how to coordinate them in cooperative activities. As they discuss their problems and work out their differences, they develop their conceptions of what is fair and just.

Whatever the interactions are specifically like, they work best, Kohlberg says, when they are open and democratic. The less children feel pressured simply to conform to authority, the freer they are to settle their own differences and formulate their own ideas. We will discuss Kohlberg's efforts to induce developmental change in the section on implications for education.

The Stage Concept

Piaget, you will recall, proposed that true mental stages meet several criteria. They (1) are qualitatively different ways of thinking, (2) are structured wholes, (3) progress in an invariant sequence, (4) can be characterized as hierarchic integrations. and (5) are cross-cultural universals. Kohlberg has taken these criteria very seriously, trying to show how his stages meet them all. Let us consider these points one at a time.

Qualitative differences

It seems fairly clear that Kohlberg's stages are qualitatively different from one another. For example, stage 1 responses, which focus on obedience to authority, sound very different from stage 2 responses, which argue that each person is free to behave as he or she wishes. The two stages do not seem to differ along any quantitative dimension, they seem qualitatively different.

Structured wholes

By "structured wholes," Kohlberg means that the stages are

not just isolated responses but are *general* patterns of thought that will consistently show up across many different kinds of issues. One gets a sense that this is true by reading through his scoring manual; one finds the same kinds of thinking reappearing on diverse items. For example, one item asks, "Why should a promise be kept?" As on the Heinz dilemma, children at stage 1 again speak in terms of obedience to rules, whereas those at stage 2 focus on exchanging favors that are in one's self-interest (e.g., "You never know when you're going to need that person to do something for you"). Similarly, as children proceed through the stages they keep giving responses that are similar to those to the Heinz dilemma.

In addition, Kohlberg and his co-workers have obtained quantitative estimates of the extent to which subjects respond in terms of one particular stage. Since some subjects might be in transition between stages, one does not expect perfect consistency. Nevertheless, Kohlberg found that subjects scored at their dominant stage across nine dilemmas about two-thirds of the time. This seems to be a fair degree of consistency, suggesting the stages may reflect general modes of thought.

Invariant sequence

Kohlberg believes that his stages unfold in an invariant sequence. Children always go from stage 1 to stage 2 to stage 3 and so forth. They do not skip stages or move through them in mixed-up orders. Not all children necessarily reach the highest stages; they might lack intellectual stimulation. But to the extent they do go through the stages, they proceed in order.

Most of Kohlberg's evidence on his stage sequence comes from *cross-sectional* data. That is, he interviewed different children at various ages to see if the younger ones were at lower stages than the older ones. Stages 1 and 2 are primarily found at the youngest age, whereas the higher stages become more prevalent as age increases. Thus, the data support the stage sequence.

Cross-sectional findings, however, are inconclusive. In a cross-sectional study, different children are interviewed at each age, so there is no guarantee that any individual child actually moves through the stages in order. For example, there is no guarantee that a boy who is coded at stage 3 at age 13 actually passed through stages 1 and 2 in order when he was younger. More conclusive evidence must come from *longitudinal* studies, in which the same children are followed over time.

The first two major longitudinal studies began with samples of teenagers and then tested them at three-year intervals. These studies produced ambiguous results. In both, most subjects either remained at the same stage or moved up one stage, but there were also some who might have skipped a stage. Furthermore, these studies indicated that some subjects had regressed, and this finding also bothered Kohlberg, because he believes that movement through his stages should always be forward.

Kohlberg's response to these troublesome findings was to revise his scoring method. He had already become uncomfortable with his first scoring manual, believing that it relied too heavily on the *content* of subjects' answers rather than their underlying *reasoning*. and he had made some improvements on it. So, when these longitudinal findings emerged, he decided to develop a much more precise and adequate scoring system and, to some extent, to revise his definitions of the stages.

To create the latest scoring manual, Kohlberg and his co-workers worked with seven boys from his original sample who had been retested every three or four years for 20 years. It was during this work that Kohlberg decided to drop stage 6. Kohlberg then examined the hypothesis of invariant sequence for 51 other boys from his original sample, who also had been retested at least twice (every three or four years) over the 20-year period. This time, Kohlberg and his colleagues found no stage-skipping, and only about 6 percent of the subjects showed signs of regressing.

Four recent longitudinal studies have obtained similar results although, two have found somewhat more regression. In general, then, the new longitudinal studies seem to support the invariant-sequence hypothesis. Kohlberg's new, longitudinal study has also changed the earlier picture of moral development in other ways. Stage 4 had become the dominant stage by age 16. In the new scoring system, however, it is more difficult to achieve the higher stages—the reasoning must be more clearly demonstrated—and Kohlberg finds that stage 4 does not become dominant until the boys are in their 20s and 30s. Stage 5, too, only appears in the mid-20s and never becomes very prevalent.

Hierarchic integration

When Kohlberg says that his stages are hierarchically integrated, he means that people do not lose the insights gained at earlier stages, but integrate them into new, broader frameworks. For example, people at stage 4 can still understand stage 3 arguments, but they now subordinate them to wider considerations. They understand that Heinz had good motives for stealing, but they point out that if we all stole whenever we had a good motive, the social structure would break down. Thus stage 4 subordinates a concern for motives to a wider concern for the society as a whole.

The concept of hierarchic integration is very important for Kohlberg because it enables him to explain the direction of his stage sequence. Since he is not a maturationist, he cannot simply say that the sequence is wired into the genes. So he wants to show how each new stage provides a broader framework for dealing with moral issues. Stage 4, as mentioned, transcends the limitations of stage 3 and becomes more broadly concerned with social organization.

Stage 5, in turn, sees the weakness of stage 4; a well-organized society is not necessarily a moral one. Stage 5 therefore considers the rights and orderly processes that make for a moral society. Each new stage retains the insights of the prior stage, but it recasts them into a broader framework. In this sense, each new stage is more cognitively adequate than the prior stage.

If Kohlberg is right about the hierarchic nature of his stages, we would expect that people would still be able to understand earlier stages but consider them inferior, In fact, when Rest presented adolescents with arguments from different stages, this is what he found. They understood lower-stage reasoning, but they disliked it. What they preferred was the highest stage they heard, whether they fully understood it or not. This finding suggests, perhaps, that they had some intuitive sense of the greater adequacy of the higher stages.

Werner, described hierarchic integration as a process that occurs alongside *differentiation,* and Kohlberg believes his stages represent increasingly differentiated structures as well. Kohlberg points out that the stage 5 value on life, for example, has become differentiated from other considerations.

Stage 5 respondents say that we ought to value life for its own sake, regardless of its value to authorities (stage 1), its usefulness to oneself (stage 2), the affection it arouses in us (stage 3), or its value within a particular social order (stage 4). Stage 5 subjects have abstracted this value from other considerations and now treat it as a purely moral ideal. Their thinking, Kohlberg says, is becoming like that of the moral philosophers in the Kantian tradition.

Universal sequence

Kohlberg, like all stage theorists, maintains that his stage sequence is universal; it is the same in all cultures. At first glance, this proposal might be surprising. Don't different cultures socialize their children differently, teaching them very different moral beliefs?

Kohlberg's response is that different cultures do teach different beliefs, but that his stages refer not to specific beliefs but to underlying modes of reasoning. For example, one culture might discourage physical fighting, while another encourages it more. As a result, children will have different beliefs about fighting, but they will still reason about it in the same way at the same stage. At stage 1, for example, one child might say that it is wrong to fight when insulted "because you will get punished for it, "while another says that "it is all right to fight; you won't get punished."

The beliefs differ, but both children reason about them in the same underlying way, in terms of the physical consequences (punishment). They do so because this is what they can cognitively grasp. Later on, the first child might argue that fighting is bad "because if everyone fought all the time there would be anarchy," while the second child argues that "people must defend their honor, because if they don't everyone will be insulting everyone, and the whole society will break down." Once again, the specific beliefs differ, reflecting different cultural teachings, but the underlying reasoning is the same—in this case it is stage 4, where people can consider something as abstract as the social order. Children, regardless of their beliefs, will always move to stage 4 thinking some time after stage 1 thinking because it is cognitively so much more sophisticated.

Kohlberg, then, proposes that his stage sequence will be the same in all cultures, for each stage is conceptually more advanced than the next. He and other researchers have given his interview to children and adults in a variety of cultures, including Mexico, Taiwan, Turkey, Israel, the Yucatan, Kenya, the Bahamas, and India. Most of the studies have been cross sectional, but a few have been longitudinal. Thus far, the studies have supported Kohlberg's stage sequence. To the extent that children move through the stages, they appear to move in order.

At the same time, people in different cultures seem to move through the sequence at different rates and to reach different end-points. In the United States most urban middle-class adults reach stage 4, with a small percentage using some stage 5 reasoning. In urban areas of other countries the picture is fairly similar. In the isolated villages and tribal communities of many countries, however, it is rare to find any adult beyond stage 3.

Kohlberg suggests that one can understand these findings in terms of Piagetian theory. Cultural factors, in this theory, do not directly shape the child's moral thought, but they do stimulate thinking. Social experiences can challenge children's ideas, motivating them to come up with new ones. In traditional villages, however, there may be little to challenge a stage 3 morality; the norms of care and empathy work very well in governing the face-to-face interactions of the group. Thus, there is little to stimulate thinking beyond this stage.

When, in contrast, young people leave the village and go off to the city, they witness the breakdown of interpersonal ties. They see that group norms of care and empathy have little impact on the impersonal interactions of city life, and they see the need for a formal legal structure to ensure moral conduct. They begin to think in terms of stage 4 morality. Furthermore, as Keniston notes, if young people attend the universities, they may take classes in which the teachers deliberately question the unexamined assumptions of their childhoods and adolescences. Thus they are stimulated to think about moral matters in new ways.

Moral Thought and Moral Behavior

Kohlberg's scale has to do with moral thinking, not moral action. As everyone knows, people who can talk at a high moral level may not behave accordingly. Consequently, we would not expect perfect correlations between moral judgment and moral action. Still, Kohlberg thinks that there should be some relationship.

As a general hypothesis, he proposes that moral behavior is more consistent, predictable. and responsible at the higher stages, because the stages themselves increasingly employ more stable and general standards. For example, whereas stage 3 bases decisions on others' feelings, which can vary, stage 4 refers to set rules and laws. Thus, we can expect that moral behavior, too, will become more consistent as people move up the sequence. Generally speaking, there is some research support for this hypothesis (e.g., with respect to cheating), but the evidence is not clear-cut.

Some research has focused on the relationships between particular stages and specific kinds of behavior. For example, one might expect that juvenile delinquents or criminals would typically reason at stages 1 or 2, viewing morality as something imposed from without (stage 1) or as a matter of self-interest (stage 2), rather than identifying with society's conventional expectations (stages 3 and 4). Again, some research supports this hypothesis, but there also are some ambiguous results.

Several studies have examined the relationship between postconventional thinking and student protest. In a landmark study, Haan et al. examined the moral reasoning of those who participated in the Berkeley Free Speech Movement in 1964. Haan found that their thinking was more strongly postconventional than that of a matched sample of nonparticipants, but this f inding was not replicated for some other protests, apparently because moral principles were not at stake.

Blasi describes, there is a relationship between moral thought and action, but he suggests that we need to introduce other variables to clarify this relationship. One variable may simply be the extent to which individuals themselves feel the need to maintain consistency between their moral thoughts and actions.

ERIK ERIKSON'S PSYCHOSOCIAL THEORY

Erik Erikson's psychosocial theory on social development is an approach to the personality that extends Freudian psychosexual theory. Erikson's theory is unique in that it encompasses the entire life cycle and recognizes the impact of society, history, and culture on personality. Erikson is best known for his concept of the identity crisis. This idea may have stemmed from a personal identity crisis he experienced at a young age. He once wrote, "No doubt my best friends will insist that I needed to name this crisis and to see it in everybody else in order to really come to terms with it myself".

Eight stages of development

Erikson's theory consist of eight stages of development. Each stage ischaracterized by a different conflict that must be resolved by the individual.When the environment makes new demands on people, the conflicts arise. "Theperson is faced with a choice between two ways of coping with each crisis, anadaptive, or maladaptive way. Only when each crisis is resolved, which involvesa change in the personality, does the person have sufficient strength to dealwith the next stages of development". If aperson is unable to resolve a conflict at a particular stage, they will confrontand struggle with it later in life.

Stage 1: Oral-Sensory

The important event in this stage is feeding. According to Erikson, theinfant will develop a sense of trust only if the parent or caregiver isresponsive and consistent with the basic needs being meet. The need for careand food must be met with comforting regularity. The infant must first form atrusting relationship with the parent or caregiver, otherwise a sense ofmistrust will develop.

The infant's need for care, familiarity, comfort and nourishment are met. Parental consistency and responsiveness is essential for the sense of trust todevelop. Babies who are not securely attached to their mothers are less cooperativeand more aggressive in their interactions with their mothers. As they growolder, they become less competent and sympathetic with peers. They also exploretheir environment with less enthusiasm and persistence. For examples: Babies will begin to understand that objects and people exist even when theycannot see them. This is where trust becomes important.

Stage 2: Muscular-Anal

According to Erikson, self control and self confidence begin to develop atthis stage. Children can do more on their own. Toilet training is the mostimportant event at this stage. They also begin to feed and dress themselves.This is how the toddler strives for autonomy. It is essential for parents not tobe overprotective at this stage. A parent's level of protectiveness willinfluence the child's ability to achieve autonomy. If a parent is notreinforcing, the child will feel shameful and will learn to doubt his or herabilities. "Erikson believes that children who experience too much doubt atthis stage will lack confidence in their powers later in life".

The child must take more responsibility for his or her own feeding,toileting, and dressing. Parents must be reassuring yet avoid overprotection. If parents do not maintain a reassuring, confident attitude and do notreinforce the child's efforts to master basic motor and cognitive skills,children may begin to feel shame; they may learn to doubt their abilities tomanage the world on their own terms. Children who experience too much doubt at this stage will lack confidence in their own powers throughout life. In this stage children begin to assume important responsibilities forself-care like feeding, toileting, and dressing.

Stage 3: Locomotor

The most important event at this stage is independence. The child continuesto be assertive and to take the initiative. Playing and hero worshipping are animportant form of initiative for children. Children in this stage are eager forresponsibility. It is essential for adults to confirm that the child'sinitiative is accepted no matter how small it may be. If the child is not givena chance to be responsible and do things on their own, a sense of guilt maydevelop. The child will come to believe that what they want to do is alwayswrong.

In order for a positive outcome in this stage, the child must learn toaccept without guilt, that there are certain things not allowed. Children mustbe guilt free when using imagination. They must be reassured that it is okay toplay certain adult roles. If children are not allowed to do things on their own, a sense of guilt maydevelop and they may come to believe that what they want to do is always wrong. A four year old passing tools to a parent who is fixing a bicycle. Childrenat this stage will worship heroes. Pretend games are also common.

Stage 4: Latency

"In this stage children are learning to see the relationship betweenperseverance and the pleasure of a job completed". The important event at this stage is attendance at school. As a student, thechildren have a need to be productive and do work on their own. They are bothphysically and mentally ready for it. Interaction with peers at school alsoplays an imperative role of child development in this stage. The child for thefirst time has a wide variety of events to deal with, including academics, groupactivities, and friends. Difficulty with any of these leads to a sense ofinferiority.

It is essential for the child at this stage to discover pleasure in beingproductive and the need to succeed. The child's relationship with peers inschool and the neighborhood become increasingly important. Difficulty with the child's ability to move between the world at home and the world of peers can lead to feeling of inferiority. In this stage children want to do productive work on their own. Students areable to water class plants, collect and distribute materials for teacher, andkeep records of forms for teacher.

Stage 5: Adolescence

At this stage, adolescents are in search of an identity that will lead themto adulthood. Adolescents make a strong effort to answer the question "Whoam I?" Erikson notes the healthy resolution of earlier conflicts can nowserve as a foundation for the search for an identity. If the child overcomesearlier conflicts they are prepared to search for identity. Did they develop thebasic sense of trust? Do they have a strong sense of industry to believe inthemselves?

The adolescent must make a conscious search for identity. This is built onthe outcome and resolution to conflict in earlier stages. If the adolescent can not make deliberate decisions and choices, especiallyabout vocation, sexual orientation, and life in general, role confusion becomesa threat. Adolescents attempt to establish their own identities and see themselves asseparate from their parents.

Stage 6: Young Adulthood

In this stage, the most important events are love relationships. Intimacyrefers to one's ability to relate to another human being on a deep, personallevel. An individual who has not developed a sense of identity usually will feara committed relationship and may retreat into isolation. It is important tomention that having a sexual relationship does not indicate intimacy. People canbe sexually intimate without being committed and open with another. Trueintimacy requires personal commitment. However, mutual satisfaction willincrease the closeness of people in a true intimate relationship.

The young adult must develop intimate relationships with others. Notresolving this conflict leaves the young adult feeling isolated. The young adultmust be willing to be open and committed to another individual. An individual may retreat into isolation if a sense of identity is notdeveloped and will fear a committed relationship.

Stage 7: Middle Adulthood

In this stage generativity refers to the adult's ability to care for anotherperson. The most important event in this stage is parenting. Does the adult havethe ability to care and guide the next generation? Generativity has a broadermeaning then just having children. Each adult must have some way to satisfy andsupport the next generation. According to Erikson, "A person does best atthis time to put aside thoughts of death and balance its certainty with the onlyhappiness that is lasting: to increase, by whatever is yours to give, the goodwill and higher order in your sector of the world".

To have and nurture children and/or become involved with future generations. An individual must deal with issues they are concerned with or it can leadto stagnation in later life. In this stage an adult will be concerned with issues such as: the future ofthe environment, what kind of world will we leave the next generation, equalityfor all people, etc.

Stage 8: Maturity

The most important event at this stage is coming to accept one's whole lifeand reflecting on that life in a positive manner. According to Erikson,achieving a sense of integrity means fully accepting oneself and coming to termswith the death. Accepting responsibility for your life and being able to undothe past and achieve satisfaction with self is essential. The inability to dothis results in a feeling of despair.

The adult feels a sense of fulfillment about life and accepts death as anunavoidable reality. Individuals who are unable to obtain a feeling of fulfillment andcompleteness will despair and fear death. An aged person may find it necessary to reflect and analyze what they haveaccumulated throughout life and decide what offspring will receive from themupon death.

Kohlberg has also tried to relate his moral stages to other forms of cognition. He has first analyzed his stages in terms of their underlying cognitive structures and has then looked for parallels in purely logical and social thought. For this purpose, he has analyzed his own stages in terms of implicit *role-taking capacities,* capacities to consider others' viewpoints. At first, at stage 1, children hardly seem to recognize that viewpoints differ. They assume that there is only one right view, that of authorities. At stage 2, in contrast, they recognize that people have different interests and viewpoints. They seem to be overcoming egocentrism; they see that perspectives are relative to the individual . They also begin to consider how individuals might coordinate their interests in terms of mutually beneficial deals.

At stage 3, people conceptualize role-taking as a deeper, more empathic process; one becomes concerned with the other's feelings. Stage 4, in turn, has a broader, society-wide conception of how people coordinate their roles through the legal system..

Stages 5 and 6, finally, take a more idealized look at how people might coordinate their interests. Stage 5 emphasizes democratic processes, and stage 6 considers how all parties take one another's perspectives according to the principles of justice.

The moral stages, then, reflect expanded insights into how perspectives differ and might be coordinated. As such, the moral stages might be related to stages of logical and social thought which contain similar insights. So far, the empirical evidence suggests that advances in moral thinking may rest upon prior achievements in these other realms. For example, children seem to advance to stage 2, overcoming their egocentrism in the moral sphere, only after they have made equivalent progress in their logical and social thought. If this pattern is correct, we can expect to find many individuals who are logical and even socially insightful but still underdeveloped in their moral judgment.

IMPLICATION FOR EDUCATION

Kohlberg would like to see people advance to the highest

possible stage of moral thought. The best possible society would contain individuals who not only understand the need for social order (stage 4) but can entertain visions of universal principles, such as justice and liberty (stage 6). How, then, can one promote moral development? Turiel found that when children listened to adults' moral judgments, the resulting change was slight. This is what Kohlberg might have expected, for he believes that if children are to reorganize their thinking, they must be more active.

Accordingly, Kohlberg encouraged another student, Moshe Blatt, to lead discussion groups in which children had a chance to grapple actively with moral issues. Blatt presented moral dilemmas which engaged the classes in a good deal of heated debate. He tried to leave much of the discussion to the children themselves, stepping in only to summarize, clarify, and sometimes present a view himself. He encouraged arguments that were one stage above those of most of the class. In essence, he tried to implement one of Kohlberg's main ideas on how children move through the stages. They do so by encountering views which challenge their thinking and stimulate them to formulate better arguments.

The Kohlberg-Blatt method of inducing cognitive conflict exemplifies Piaget's equilibration model. The child takes one view, becomes confused by discrepant information, and then resolves the confusion by forming a more advanced and comprehensive position. The method is also the dialectic process of Socratic teaching. The students give a view, the teacher asks questions which get them to see the inadequacies of their views, and they are then motivated to formulate better positions.

In Blatt's first experiment, the students (sixth graders) participated in 12 weekly discussion groups. Blatt found that over half the students moved up one full stage after the 12 weeks. Blatt and others have tried to replicate these findings, sometimes using other age groups and lengthier series of classes. As often happens with replications, the results have not been quite so successful; upward changes have been smaller—usually a third of a stage or less, Still, it generally seems that Socratic classroom discussions held over several months can produce changes that, while small, are significantly greater than those found in control groups who do not receive these experiences.

One of Blatt's supplementary findings was that those students who reported that they were most "interested" in the discussions made the greatest amount of change. This finding is in keeping with Piagetian theory. Children develop not because they are shaped through external reinforcements but because their curiosity is aroused. They become interested in information that does not quite fit into their existing cognitive structures and are thereby motivated to revise their thinking Another Kohlberg student—M. Berkowitz (1980)—is examining actual dialogues to see if those who become most challenged and involved in the tensions of moral debate are also those who move forward.

Although Kohlberg remains committed to the cognitive-conflict model of change, he has also become interested in other strategies. One is the "just Community" approach. Here the focus is not the individuals but groups. For example, Kohlberg and some of his colleagues set up a special democratic high school group and actively encouraged the students to think of themselves as a community. Initially, little community feeling was present. The group's dominant orientation was stage 2; it treated problems such as stealing as purely individual matters. If a boy had something stolen, it was too bad for him. After a year, however, the group norms advanced to stage 3; the students now considered stealing to be a community issue that reflected on the degree of trust and care in the group.

It will be interesting to see if the just community approach can promote further advances in moral thinking. In the meantime, this approach has aroused some uneasiness among some of Kohlberg's associates. In particular, Reimer et al. have wondered whether Kohlberg, by explicitly encouraging the students to think of themselves as a community, is not practicing a form of indoctrination. Reimer says that he has talked to Kohlberg about this, and he has come away convinced that Kohlberg is committed to democratic groups in which students are encouraged "to think critically, to discuss assumptions, and. when they feel it is necessary, to challenge the teacher's suggestions". Thus, moral development remains a product of the students' own thinking.

2

Principles of Growth and Development

There is a set of principles that characterizes the pattern and process of growth and development. These principles or characteristics describe typical development as a predictable and orderly process; that is, we can predict how most children will develop and that they will develop at the same rate and at about the same time as other children. Although there are individual differences in children's personalities, activity levels, and timing of developmental milestones, such as ages and stages, the principles and characteristics of development are universal patterns.

PRINCIPLES OF DEVELOPMENT

Development proceeds from the head downward. This is called the cephalocaudle principle. This principle describes the direction of growth and development. According to this principle, the child gains control of the head first, then the arms, and then the legs. Infants develop control of the head and face movements within the first two months after birth. In the next few months, they are able to lift themselves up by using their arms. By 6 to 12 months of age, infants start to gain leg control and may be able to crawl, stand, or walk. Coordination of arms always precedes coordination of legs.

Development proceeds from the center of the body outward. This is the principle of proximodistal development that also describes the direction of development. This means that the spinal cord develops before outer parts of the body. The child's arms develop before the hands and the hands and feet develop before the fingers and toes. Finger and toe muscles (used in fine motor dexterity) are the last to develop in physical development.

Development depends on maturation and learning. Maturation refers to the sequential characteristic of biological growth and development. The biological changes occur in sequential order and give children new abilities. Changes in the brain and nervous system account largely for maturation. These changes in the brain and nervous system help children to improve in thinking (cognitive) and motor (physical) skills. Also, children must mature to a certain point before they can progress to new skills (Readiness).

For example, a four-month-old cannot use language because the infant's brain has not matured enough to allow the child to talk. By two years old, the brain has developed further and with help from others, the child will have the capacity to say and understand words. Also, a child can't write or draw until he has developed the motor control to hold a pencil or crayon. Maturational patterns are innate, that is, genetically programmed. The child's environment and the learning that occurs as a result of the child's experiences largely determine whether the child will reach optimal development. A stimulating environment and varied experiences allow a child to develop to his or her potential.

Development proceeds from the simple (concrete) to the more complex. Children use their cognitive and language skills to reason and solve problems. For example, learning relationships between things (how things are similar), or classification, is an important ability in cognitive development. The cognitive process of learning how an apple and orange are alike begins with the most simplistic or concrete thought of describing the two. Seeing no relationship, a preschool child will describe the objects according to some property of the object, such as color. Such a response would be, "An apple is red (or green) and an orange is orange."

The first level of thinking about how objects are alike is to give a description or functional relationship (both concrete thoughts) between the two objects. "An apple and orange are round" and "An apple and orange are alike because you eat them" are typical responses of three, four and five year olds. As children develop further in cognitive skills, they are able to understand a higher and more complex relationship between objects and things; that is, that an apple and orange exist in a class called fruit. The child cognitively is then capable of classification.

Growth and development is a continuous process. As a child develops, he or she adds to the skills already acquired and the new skills become the basis for further achievement and mastery of skills. Most children follow a similar pattern. Also, one stage of development lays the foundation for the next stage of development. For example, in motor development, there is a predictable sequence of developments that occur before walking. The infant lifts and turns the head before he or she can turn over.

Infants can move their limbs (arms and legs) before grasping an object. Mastery of climbing stairs involves increasing skills from holding on to walking alone. By the age of four, most children can walk up and down stairs with alternating feet. As in maturation, in order for children to write or draw, they must have developed the manual (hand) control to hold a pencil and crayon.

Growth and development proceed from the general to specific. In motor development, the infant will be able to grasp an object with the whole hand before using only the thumb and forefinger. The infant's first motor movements are very generalized, undirected, and reflexive, waving arms or kicking before being able to reach or creep toward an object. Growth occurs from large muscle movements to more refined (smaller) muscle movements.

There are individual rates of growth and development. Each child is different and the rates at which individual children grow is different. Although the patterns and sequences for growth and development are usually the same for all children, the rates at which individual children reach developmental stages will be different. Understanding this fact of individual differences in rates of development should cause us to be careful about using and relying on age and stage characteristics to describe or label children. There is a range of ages for any developmental task to take place. This dismisses the notion of the "average child".

Some children will walk at ten months while others walk a few months older at eighteen months of age. Some children are more active while others are more passive. This does not mean that the passive child will be less intelligent as an adult. There is no validity to comparing one child's progress with or against another child. Rates of development also are not uniform within an individual child. For example, a child's intellectual development may progress faster than his emotional or social development.

An understanding of the principles of development helps us to plan appropriate activities and stimulating and enriching experiences for children, and provides a basis for understanding how to encourage and support young children's learning. These principles or characteristics describe typical development as a predictable and orderly process; that is, we can predict how most children will develop and that they will develop at the same rate and at about the same time as other children. Although there are individual differences in children's personalities, activity levels, and timing of developmental milestones, such as ages and stages, the principles and characteristics of development are universal patterns.

PRINCIPLES OF CHILD DEVELOPMENT

A Summary Experiences for children, at whatever age, should be designed around the child's developmental needs and abilities. Support for this position comes from international literature on how children develop and learn.

Development is holistic; it consists of inter-dependent dimensions. This means that the child's development cannot be compartmentalized into health, nutrition, education, social, emotional and spiritual variables. All are interwoven in a child's life and are developing simultaneously. Progress in one area affects progress in others. Similarly, when something goes wrong in any one of those areas, it has an impact on all the other areas.

Development begins prenatally and learning is occurring at birth.

From the moment of conception important developments occur that affect the brain, the physical body, and the chemistry of the child. These all have an impact on that child's ability to learn, to thrive, to grow and be healthy.

The first 8 years of a child's life form the foundation for all later development; early attention to the child's needs is critical. Developmental psychologists have demonstrated that in the early years a child develops all the basic brain and physiological structures upon which later growth and learning are dependent. The benefits of intervention are greater if you provide appropriate supports early. It is also important that there be continuous attention to the child's development. Good care for infants (health, nutritious foods, active feeding, mental stimulation and interaction) is the best preventive measure to avoid disabilities and developmental delays.

Children's needs differ across the early childhood years.

A child's developmental needs will vary over the first years of life in relation to his or her maturation process. The following periods can be distinguished:

Prenatal and birth. During this period, attention is focused on the mother through maternal and child health programs, and parental support and education.

Infancy (up to about 18 months). Particular emphasis needs to be placed on health and nutrition during this period. Breastfeeding is critical; every attempt should be made for children to be with their mothers. Sensory learning, especially auditory and visual, control of physical actions, and attachment to a significant caregiver seem to be the central tasks.

Toddler and post-toddler (about 18 to 36 months). During this period nutrition continues to be critically important. A child's coordination, language, ability to think and social skills advance rapidly.

Pre-school (approximately ages 3, 4, 5 and sometimes 6). In the pre-school years, socialization and preparation for schooling take on greater importance, and the circle of peers and caregivers widens.

From age four onward, early childhood programs are more likely to be associated with education and preschools, but health and nutrition remain key components of what young children need.

Early primary school, a period of "transition" into school and the world at large (roughly ages 6 to 8). This transition can be relatively easy or extremely difficult. In either case children continue to need holistic attention.

Development is multi-determined, and varies as a function of the child's nutritional and biomedical status, genetic inheritance, and social and cultural context. While there is a general pattern or sequence for development that is true of most children, the rate, character, and quality of development vary from child to child as a result of what the child brings into the world (the child's nature) and the experiences the child has in the world (nurture). Culture influences development in different ways. It is important for adults to use methods that fit with the child's growth pattern, not only in the cognitive (intellectual) area, but also in the emotional, perceptual and motor areas.

Children's development is cumulative in nature. If children's early nutrition and health are jeopardized, development is delayed or debilitated, and, over time the child's development follows a downward trajectory. As a result, some children fall further and further behind their peers. The earlier attention is given to children's health, nutritional and developmental status, the better. If interventions are going to be provided only after the age of three, there may be limits on what can be achieved in terms of overcoming already existing deficits resulting from poor nutrition, illness and neglect. If adequate care (particularly in relation to nutrition, feeding and stimulation) is provided in the first three years, then there can be some very positive long-term outcomes from programs which focus on the pre-school years.

Children are active participants in their own development and learning.

Learning involves the child's construction of knowledge, not an adult's imposition of information. It is important for children to have opportunities to construct their own knowledge through exploration, interaction with materials and imitation of role models. Children need opportunities to learn by doing, to be engaged in problem-solving, and to develop language and communication skills. Opportunities for active involvement should abound, whether at home in everyday chores or in more organized settings outside the home. Emphasis should be on how to learn (i.e., positive attitudes toward learning and thinking skills) rather than on what to learn.

Development and learning occur as a result of the child interacting with people and objects in his or her environment.

Children need opportunities to interact with peers, adults and objects in a safe environment that provides the child with security and acceptance. Learning and development can be fostered by creating a healthy environment, and providing space, materials, and opportunities to help children to learn through play, whether at home or elsewhere.

Children live within a context–family, community, culture–and their needs are most effectively addressed in relation to that context. The child's well-being is closely linked to the well-being of the family, specifically to the well-being of the primary caregiver(s). Therefore, support to the family and community can help children; similarly support to children can help the family and community. Since the environment has an impact on children's development, it is also possible to develop interventions that make changes in the child's environment.

THEORETICAL PRINCIPLES OF CHILD DEVELOPMENT AND LEARNING

The following are theoretical principles of child development and learning that are critical in developmentally appropriate practice (DAP). These principles are based on the work of Piaget, Vygotsky, Erikson, and others.

Children learn best when their physical needs are met and they feel psychologically safe and secure.

DAP respects children's biological needs. For example, children are not made to sit and attend to paperwork or listen to adult lectures for long periods of time. DAP calls for active play and periods of quiet, restful, activity. The environment is safe and secure where everyone is accepted.

Children construct knowledge.

Knowledge is constructed as a result of dynamic interactions between the individual and the physical and social environments. In a sense the child discovers knowledge through active experimentation. Central to experimentation is making "constructive errors" that are necessary to mental development. Children need to form their own hypotheses and keep trying them out through mental actions and physical manipulations - observing what happens, comparing their findings, asking questions, and discovering answers - and adjust the model or alter the mental structures to account for the new information.

Children learn through social interaction with other adults and other children.

A prime example is the parent-child relationship. The teacher encourages and fosters this relationship as well as relationships with peers and other adults by supporting the child in his or her efforts and later allowing the child to function independently. The teacher's role is one of supporting, guiding, and facilitating development and learning.

Children learn through play.

Play provides opportunities for exploration, experimentation, and manipulation that are essential for constructing knowledge and contributes to the development of representational thought. During play, children examine and refine their learning in light of the feedback they receive from the environment and other people. It is through play that children develop their imaginations and creativity. During the primary grades, children's play becomes more rule-oriented and promotes the development of autonomy and cooperation which contributes to social, emotional, and intellectual development.

Children's interests and "need to know" motivate learning.

Children have a need to make sense of their experiences. In a developmentally appropriate classroom, teachers identify what intrigues their children and then allow the students to solve problems together. Activities that are based on children's interests provide motivation for learning. This fosters a love of learning, curiosity, attention, and self-direction.

Human development and learning and are characterized by individual variation.

A wide range of individual variation is normal and to be expected. Each human being has an individual pattern and timing of growth development as well as individual styles of learning. Personal family experiences and cultural backgrounds also vary.

EARLY CHILDHOOD GUIDELINES

An important contribution to the field of child development and early childhood education was the creation of "Guidelines for Developmentally Appropriate Curriculum and Assessment in Programs Serving Children 3 through 8". In US, the National Association for the Education of Young Children (NAEYC) and the National Association of Early Childhood Specialists in State Departments of Education (NAECS/SDE) jointly developed these guidelines to assist teachers and supervisors to:

1) Make informed decisions about appropriate curriculum, content, and assessment;
2) Evaluate existing curriculum and assessment practices; and,
3) Advocate for more appropriate approaches.

The early childhood profession believes that curriculum and assessment should be based on the best knowledge of theory and research about how children develop and learn with attention given to individual children's needs and interests in a group in relation to program goals. It is important, therefore, to clarify the definitions of these important terms.

Curriculum is an organized framework that delineates the content children are to learn, the processes through which children achieve the identified curricular goals, what teachers do to help children achieve these goals, and the context in which teaching and learning occur. Assessment is the ongoing process of observing, recording and otherwise documenting the work children do and how they do it, to provide a basis for a variety of educational decisions that affect the child.

Assessment is integral to curriculum and instruction. In early childhood programs, assessment provides a basis for: 1) planning instruction and communicating with parents; 2) identifying children with special needs; and 3) evaluating programs and demonstrating accountability.

Determining Appropriate Content

Learning and development are so individualized, it is not possible nor desirable to establish uniform age-appropriate expectations. However, it is possible to identify parameters to guide decisions about the appropriateness of curriculum expectations.

The framework that follows is useful for determining age-appropriate curriculum content. This framework reflects the cycle of human learning—movement from awareness, to exploration, to inquiry, to utilization.

- Awareness is broad recognition of the parameters of the learning—events, objects, people, or concepts.
- Exploration is the process of figuring out the components or attributes of events, objects, people, or concepts by whatever means available; it also is the process whereby children bring their own personal meaning to their experiences.
- Inquiry is the process of developing understanding of commonalities across events, objects, people, or concepts. At this point, children begin to generalize their personal concepts and adapt them to more adult ways of thinking and behaving.

— Utilization is the functional level of learning, at which children can apply or make use of their understanding of events, objects, people, or concepts.

To learn something new, children must become aware, explore, inquire, use, and apply. This process occurs over time and reflects movement from learning that is informal and incidental, spontaneous, concrete-referenced, and governed by the child's own rules to learning that is more formal, refined, extended, enriched, more removed in time and space from concrete references and more reflective of conventional rule systems.

Conditions Necessary for Change

Experts agree that meaningful change must be systemic. In other words, change must occur in all aspects and levels of the educational system. This system extends far beyond the walls of the early childhood classroom, into the school building, the community, nation, and global society. The term systemic also suggests that change in one aspect of the educational system will affect other aspects as well.

To complicate matters, each of these aspects also may be in the process of change. Thus in the change process, all those involved must continually communicate and take note of the whole educational system, evaluating current plans within the context of the changing whole. We have identified several important conditions necessary for systemic change in early childhood education.

1. We must focus on how children develop and learn in order to meet their cognitive, social, emotional, and physical needs. This calls for early childhood educators to have a thorough understanding of child development, and be able to implement that knowledge in all aspects of their classrooms.

2. The primary school must have the support of the entire community, including the elementary and secondary school

staff. Developmentally appropriate practice includes smooth transitions from one grade to the next, so teachers must work together to insure common curriculum and assessment practices that provide continuity throughout the child's academic career.

3. Parents must be encouraged to take active roles in the education of their children. Measures must be taken to involve them in all aspects of their child's growth and development in school, allowing them to be partners in their child's education. This necessitates open communication, problem-solving, and collaborative decisionmaking between and among administrators, teachers and parents.

ACTIVITIES FOR TEACHERS

The examples of excellence in this program clearly show that in successful schools, teaching is a multidimensional activity. One of the most powerful of these dimensions is that of "teacher as researcher." Not only do teachers need to use research in their practice, they need to participate in "action" research in which they are always engaging in investigation and striving for improved learning. The key to action research is to pose a question or goal, and then design actions and evaluate progress in a systematic, cyclical fashion as the means are carried out. Below are three major ways that you can become involved as an action researcher.

1. Use the questions found at the end of this section to evaluate your school's early childhood curriculum and assessment.
2. Implement the models of excellence presented in this program.

The following are activities that individuals and groups can do with early childhood programs.

1. Serve on early childhood committees to identify early childhood needs and resources and ways you and your

group or agency can support developmentally appropriate practice in early childhood.

2. Consider ways that schools and community members can work together to provide:

— Materials for a rich early childhood learning environment (e.g., real literature in print and audio form, computers, manipulatives for learning)

— Opportunities for children and teachers to learn out of school

— Opportunities for adults to serve as role models, tutors, aides, and experts

— Opportunities for children to become involved with and contribute to the community

3. Promote school and community forums to debate the national education goals:

— Invite your local television and radio stations to host school and community forums.

— Have "revolving school/community breakfasts" (community members visit schools for breakfast once or twice a month, changing the staff and community members each time).

— Gather information on the national education goals and their assessment.

— Gather information on alternative models of schooling.

— Gather information on best practices and research in the classroom.

Some of the important questions and issues to discuss in your forums are:

— Have we reviewed the national education goals documents to arrive at a common understanding of each goal?

— What will students be like who learn in schools that achieve the goals?

— What must schools and communities be like to achieve the goals?

— Do we agree with the goals, and how high do we rate each?

— What is the reason for national pessimism about their achievement?

— How are our schools doing now in terms of achieving each?

— Why is it important for us to achieve the goals?

— What are the consequences for our community if we don't achieve them?

— What assumptions are we making about the future in terms of Knowledge, Technology and Science, Humanities, Family, Change, Population, Minority Groups, Ecology, Jobs, Global Society, and Social Responsibility? Discuss in terms of each of the goal areas.

PRINCIPLES OF SMART GROWTH

For a community to be successful in implementing smart growth, it must be embraced by the private sector. Only private capital markets can supply the large amounts of money needed to meet the growing demand for smart growth developments. If investors, bankers, developers, builders and others do not earn a profit, few smart growth projects will be built. Fortunately, government can help make smart growth profitable to private investors and developers. Since the development industry is highly regulated, the value of property and the desirability of a place is largely affected by government investment in infrastructure and government regulation. Governments that make the right infrastructure and regulatory decisions will create fair, predictable and cost effective smart growth.

Despite regulatory and financial barriers, developers have been successful in creating examples of smart growth. The process to do so, however, requires them to get variances to the codes – often a time-consuming, and therefore costly, requirement. Expediting the approval process is of particular importance for developers, for whom the common mantra, "time is money" very aptly applies. The longer it takes to get approval for building, the longer the developer's capital remains tied up in the land and not earning income.

For smart growth to flourish, state and local governments must make an effort to make development decisions about smart growth more timely, cost-effective, and predictable for developers. By creating a fertile environment for innovative, pedestrian-oriented, mixed-use projects, government can provide leadership for smart growth that the private sector is sure to support.

"Smart growth" is designed to promote development while protecting the environment, encouraging social and economic equity, and conserving energy and water resources. Smart growth will refocus a larger share of regional growth within central cities, urbanized areas, near transportation nodes, and in communities already served by infrastructure. It will reduce the share of regional growth that occurs on newly urbanizing land, farms, and environmentally sensitive areas. It will encourage more density in some places, to save precious land in other places.

Following smart growth principles:

1. Encourage community & stakeholder collaboration in development decisions.

Smart growth should respond to a community's own sense of how and where it wants to grow. The needs of every community and the programs to address them are best defined by the people who live and work there. An open public process that facilitates the participation of community residents and organizations provides community members with a direct stake in ensuring the success of revitalization efforts, and can lead to creative resolution of development issues.

2. *Integrate people and place.*

Smart growth must integrate people-focused strategies-efforts that support community residents and families-with place-focused strategies-those that support physical development and stabilize the community environment. This integrated approach is necessary to maximize community impact and to reduce unintended negative consequences to either the community or the environment.

3. *Promote regional equity and reduce local and regional disparities.*

Everyone wants to live in a community where the quality of life is high. If we cooperate across communities and plan carefully for the future, we can achieve a situation where neighborhoods share fairly in the benefits of development and none bears an unfair burden of the social costs associated with planning decisions. Metropolitan regions and communities that reduce local and regional disparities are more likely to compete successfully for national and international economic opportunities.

4. *Strengthen regional cooperation.*

Most community assets and problems do not recognize municipal boundaries. Regional cooperation is needed to address common concerns, such as traffic, and to protect common resources, such as watersheds. Regional solutions can also encourage the efficient use of funds. Smart growth encourages communities to work together collaboratively to enhance their assets as well as to solve their problems.

5. Promote distinctive, attractive communities with a strong sense of place.

Smart growth promotes development that respects and enhances natural and built features and landmarks to create a sense of defined neighborhoods, towns, and regions. It fosters physical environments that reflect the culture and values of the people who live there, while supporting a more cohesive community fabric. It encourages localities to plan wisely for the future, and gives them the power and resources to implement those plans over time.

6. Preserve open space, farmland, and critical environmental resources.

Open space, both within and surrounding developed areas, enhances quality of life by providing community space, recreation opportunities, critical habitat for plants and wildlife, working lands for farming and forestry, and preserves the quality of critical environmental resources such as wetlands, watersheds, and drinking water supplies.

7. Encourage development in currently developed areas to take advantage of existing community assets.

Smart growth encourages development in areas where public investments have already been made in infrastructure, parks, schools, and other facilities, seeking to utilize these resources and to conserve open spaces and natural resources in undeveloped areas. In cities, this can mean focusing development on infill sites and abandoned brownfields; in towns, this can mean clustering development around town centers and transportation nodes. Recognizing that the extension of roadways, sewer lines, and other forms of infrastructure can often "drive" sprawl, government agencies should coordinate decisions about public investment in infrastructure in ways that prevent this unintended consequence and encourage smart growth.

8. Mix land uses.

Smart growth supports the integration of mixed land uses (such as residential, commercial, and civic uses) into neighborhoods, communities and the region as critical components of achieving better places to live and work. By locating a variety of uses in close proximity to each other at an appropriate scale, alternatives to driving, such as walking, bicycling and transit become viable. Diverse, well designed, mixed use development also enhances the vitality and quality of life in our communities and the region as a whole, eventually weaving a fabric that supports broader housing, economic development and transportation goals.

9. Take advantage of compact development design and create walkable neighborhoods.

Smart growth provides a means for communities to incorporate more compact development design as an alternative to conventional, land consumptive development. This permits more open space to be preserved, which also reduces the impacts of storm water runoff and flooding. By drawing uses into closer proximity to each other, compact development is essential to support broader transportation choices such as transit, bicycling and walking. It also reduces the cost of providing and maintaining services such as such as water, sewer, utilities and communications.

10. Promote economic development in ways that produce jobs, strengthen low and moderate-income communities, and protect the natural environment.

Economic development is a critical objective for our region's future. Families and individuals throughout Greater Boston need new jobs and opportunities for advancement in order to enhance their futures and to make the region stronger and more competitive. Good planning practice encourages economic development in ways that minimize disruption of the natural environment. We should also strive to direct a reasonable proportion of such development to areas that presently suffer from poverty, a weak job market, and abandoned or undeveloped "brownfield" sites. Programs that provide adequate training and retraining for workers will also help to encourage employers to locate or expand in Greater Boston.

11. Create a range of housing opportunities and choices in cities and towns throughout the region.

Promote diverse housing types in all communities to enable persons and households from a wide range of economic levels, cultures, and age groups to live and work within their boundaries. No single type of housing can serve the varied needs of today's diverse households. Smart growth represents an opportunity for communities to increase housing choice not only by modifying land use patterns on newly-developed land, but also by increasing housing supply in existing neighborhoods already served by infrastructure.

12. Promote more transportation choices through the appropriate development of land.

The ability to provide people with more choices in transportation is a key element of smart growth. Better connections between appropriate land use and transportation can support the provision of alternatives to auto use, including transit, para-transit, walking, and bicycling and serve as a guide for effective transportation investments.

13. Develop predictable, fair and cost effective regulatory approvals for smart growth oriented developments.

Developments that adhere to smart growth guidelines should be allowed to proceed through public review and regulatory evaluation with a minimum of delay and cost. This will encourage developers to "build smart" while it enhances the relationship among developers, residents, and government agencies.

14. *Encourage fiscal policies that support smart growth.*

Federal, state and local \tax policies should encourage communities to base development decisions on sound planning principles, not tax benefits. Smart growth is best supported by tax policies that encourage careful planning and inter-local cooperation, and discourage "fiscal zoning," interlocal competition, and sprawl. Since sprawl often results in expenses that must be borne by the entire Commonwealth, the formula for the distribution of local aid, the "new growth" provisions of Proposition 2½, and other tools should be examined for mechanisms that might encourage smart growth and discourage sprawl.

15. *Enable smart growth by reforming existing zoning.*

Many of the principles of smart growth, such as mixed uses and higher densities, are difficult or impossible to achieve in Massachusetts under the state's zoning enabling law and current local zoning codes. For smart growth to succeed, reforms to zoning are needed to remove disincentives to smart growth.

3

VARIOUS STAGES OF DEVELOPMENT

BASIC STAGES OF CHILD DEVELOPMENT

Prenatal Development

Prenatal development is divided into three trimesters. During the first two months the developing human is referred to as an embryo. The embryo has three layers from which all body organs develop. During the second trimester the developing human is referred to as a fetus. During the third trimester, the individual is a baby which if born prematurely could survive with extra support. Premature births prior to the third trimester are less likely to survive, even with extraordinary medical care in a neonatal intensive care unit.

First Trimester

First Month

Fertilization, descent of ovum from tube to uterus. Early cell division and formation of embryonic disc from which new organism will develop. Early formation of three layers of cells: (1) the ectoderm, from which sense organs and nervous system will develop (2) the mesoderm, from which circulatory, skeletal and muscular systems will develop (3) the endoderm, from which digestive and some glandular systems will develop. Special layer of cells formed in the uterus which will become the placenta and through which nutritive substances will be carried to the new organism and waste products carried away.

Special layer of cells forms the amnion or water-sac, which will surround the developing embryo except at umbilical cord. Heart tube forms and begins to pulsate and force blood to circulate through blood vessels in embryonic disc. Nervous system begins to arise, first in form of neural groove. Development of intestinal tract, lungs, liver and kidneys begins. By end of one month, the embryo is about one-fourth inch long, curled into a crescent, with small nubbins on sides of body indicating incipient arms and legs.

Second Month

Embryo increases in size to about 1½ inches. Bones and muscle begin to round out contours of body. Face and neck develop and begin to give features a human appearance. Forehead very prominent, reflecting precocious development of brain in comparison to rest of body. Limb buds elongate. Muscles and cartilage develop. Sex organs begin to form.

Third Month

Beginning of fetal period. Sexual differentiation continues, with male sexual organs showing more rapid development and the female remaining more neutral. buds for all 20 temporary teeth laid down. Vocal cords appear; digestive system shows activity. Stomach cells begin to secrete fluid; liver pours bile into intestine. Kidneys begin functioning, with urine gradually seeping into amniotic fluid. Other waste products passed through placenta into mother's blood. Bones and muscles continue development, and by end of third month spontaneous movements of arms, legs, shoulders and fingers are possible.

Second Trimester

Fourth Month

Lower parts of body show relatively accelerated rate, so that head size decreases from one-half to one-fourth of body size. Back straightens, hands and feet are well-formed. Skin appears dark red, owing to coursing of blood showing through thin skin and wrinkles, owing to absence of underlying fat. Finger closure is possible. Reflexes become more active as muscular maturation continues. Fetus begins to stir and so thrust out arms and legs in movements readily perceived by the mother.

Fifth Month

Skin structures begin to attain final form. Sweat and sebaceous glands are formed and function. Skin derivatives also appear — hair, nails on fingers and toes. Bony axis becomes quite straight and much spontaneous activity occurs. Fetus is lean and wrinkled, about one foot long and weighs about one pound.

Sixth Month

Eyelids which have been fused shut since third month, reopen; eyes are completely formed. Taste buds appear on tongue and in mouth and are, in fact, more abundant than in the infant or adult.

Third Trimester

Seventh Month

Organism capable of independent life from this time on. Cerebral hemispheres cover almost the entire brain. Seven-month fetus can emit a variety of specialized responses.

Generally is about 15 inches long and weighs about three pounds.

Eighth and Ninth Month

During this time, finishing touches are being put on the various organs and functional capacities. Fat is formed rapidly over the entire body, smoothing out the wrinkled skin and rounding out body contours. Dull red color of skin fades so that a firth pigmentation of skin is usually very slight in all races. Activity is usually great and he can change his position within the somewhat crowded uterus. Periods of activity will alternate with periods of quiescence. Fetal organs step up their activity. Fetal heart rate becomes quite rapid. Digestive organs continue to expel more waste products, leading to the formation of a fetal stool, called the meconium, which is expelled shortly after birth. Violent uterine contractions begin, though milder ones have been tolerated earlier, and the fetus is eventually expelled from the womb into an independent physiological existence.

NORMAL STAGES OF HUMAN DEVELOPMENT (BIRTH TO 5 YEARS)

It is important to keep in mind that the time frames presented are averages and some children may achieve various developmental milestones earlier or later than the average but still be within the normal range. This information is presented to help parents understand what to expect from their child. Any questions you may have about your child's development should be shared with his doctor.

General Developmental Sequence Toddler through Preschool

This section presents typical activities and achievements for children from two to five years of age. It is important to keep in mind that the time frames presented are averages and some children may achieve various developmental milestones earlier or later than the average but still be within the normal range. This information is presented to help parents understand what to expect from their child. Any questions you may have about your child's development should be shared with his doctor or teacher.

Physical Development: Walks well, goes up and down steps alone, runs, seats self on chair, becoming independent in toileting, uses spoon and fork, imitates circular stroke, turns pages singly, kicks ball, attempts to dress self, builds tower of six cubes.

Emotional Development: Very Self-centered, just beginning a sense of personal identity and belongings, possessive, often negative, often frustrated, no ability to choose between alternatives, enjoys physical affection, resistive to change, becoming independent, more responsive to humor and distraction than discipline or reason.

Social Development: Solitary play, dependent on adult guidance, plays with dolls, refers to self by name, socially very immature, little concept of others as "people." May respond to simple direction.

Intellectual Development: Says words, phrases and simple sentences, 272 words, understands simple directions, identifies simple pictures, likes to look at books, short attention span, avoids simple hazards, can do simple form board.

Age 2

Physical Development: Runs well, marches, stands on one foot briefly, rides tricycle, imitates cross, feeds self well, puts on shoes and stockings, unbuttons and buttons, build tower of 10 cubes. Pours from pitcher.

Emotional Development: Likes to conform, easy going attitude, not so resistive to change, more secure, greater sense of personal identity, beginning to be adventuresome, enjoys music.

Age 3

Social Development: Parallel play, enjoys being by others, takes turns, knows if he is a boy or girl, enjoys brief group activities requiring no skill, likes to "help" in small ways—responds to verbal guidance.

Intellectual Development: Says short sentences, 896 words, great growth in communication, tells simple stories, uses words as tools of thought, wants to understand environment, answers questions, imaginative, may recite few nursery rhymes

Physical Development: Skips on one foot, draws "Man", cuts with scissors (not well), can wash and dry face, dress self except ties, standing broad jump, throws ball overhand, high motor drive.

Emotional Development: Seems sure of himself, out-of bounds behavior, often negative, may be defiant, seems to be testing himself out, needs controlled freedom.

Social Development: Cooperative play, enjoys other children's company, highly social, may play loosely organized group games - tag, duck-duck-goose, talkative, versatile.

Intellectual Development: Uses complete sentences, 1540 words, asks endless questions, learning to generalize, highly imaginative, dramatic, can draw recognizable simple objects.

Age 4

Physical Development: Hops and skips, dresses without help, good balance and smoother muscle action, skates, rides wagon and scooter, prints simple letters, handedness established, ties shoes, girls small muscle development about 1 year ahead of boys.

Emotional Development: Self-assured, stable, well-adjusted, home-centered, likes to associate with mother, capable, of some self-criticism, enjoys responsibility. Likes to follow the rules.

Age 5

Social Development: Highly cooperative play, has special "friends", highly organized, enjoys simple table games requiring turns and observing rules, "school", feels pride clothes and accomplishments, eager to carry out some responsibility.

Intellectual Development: 2,072 words, tells long tales, carries out direction well, reads own name, counts to 10, asks meaning of words, knows colors, beginning to know difference between fact and fiction-lying, interested in environment, city, stores, etc.

STAGES OF INTELLECTUAL DEVELOPMENT IN CHILDREN AND TEENAGERS.

Children are not little adults. Until they reach the age of 15 or so they are not capable of reasoning as an adult. The following information is based on the work of Jean Piaget. He was not a psychologist. He was a developmental biologist who devoted his life to closely observing and recording the intellectual abilities of infants, children and adolescents.

The stages of intellectual development formulated by Piaget appear to be related to major developments in brain growth. The human brain is not fully developed until late adolescence or in the case of males sometimes early adulthood. We often expect children to think like adults when they are not yet capable of doing so. It is important that parents know what to expect from their child as they develop and to be sure that the expectations they may have for their child at a given age are realistic.

PIAGET'S STAGES OF COGNITIVE DEVELOPMENT

Developmental Stage Characteristic Behavior

Reflexive Stage (0-2 months): Simple reflex activity such as grasping, sucking.

Primary Circular Reactions (2-4 months): Reflexive behaviors occur in stereotyped repetition such as opening and closing fingers repetitively.

Secondary Circular Reactions (4-8 months): Repetition of change actions to reproduce interesting consequences such as kicking one's feet to more a mobile suspended over the crib.

Coordination of Secondary Reactions (8-12 months): Responses become coordinated into more complex sequences. Actions take on an "intentional" character such as the infant reaches behind a screen to obtain a hidden object.

Tertiary Circular Reactions (12-18 months): Discovery of new ways to produce the same consequence or obtain the same goal such as the infant may pull a pillow toward him in an attempt to get a toy resting on it.

Invention of New Means Through Mental Combination (18-24 months): Evidence of an internal representational system. Symbolizing the problem-solving sequence before actually responding. Deferred imitation.

Preoperational Phase (2-4 years): Increased use of verbal representation but speech is egocentric. The beginnings of symbolic rather than simple motor play. Transductive reasoning. Can think about something without the object being present by use of language.

Intuitive Phase: (4-7 years) Speech becomes more social, less egocentric. The child has an intuitive grasp of logical concepts in some areas. However, there is still a tendency to focus attention on one aspect of an object while ignoring others. Concepts formed are crude and irreversible. Easy to believe in magical increase, decrease, disappearance. Reality not firm. Perceptions dominate judgment. In moral-ethical realm, the child is not able to show principles underlying best behavior. Rules of a game not develop, only uses simple do's and don'ts imposed by authority.

Evidence for organized,. logical thought. There is the ability to perform multiple classification tasks, order objects in a logical sequence, and comprehend the principle of conservation. thinking becomes less transductive and less egocentric. The child is capable of concrete problem-solving.

Some reversibility now possible (quantities moved can be restored such as in arithmetic: 3+4 = 7 and 7-4 = 3, etc.)

Class logic-finding bases to sort unlike objects into logical groups where previously it was on superficial perceived attribute such as color. Categorical labels such as "number" or animal" now available.

Period of Formal Operations (11-15 years)

Thought becomes more abstract, incorporating the principles of formal logic. The ability to generate abstract propositions, multiple hypotheses and their possible outcomes is evident. Thinking becomes less tied to concrete reality. Formal logical systems can be acquired. Can handle proportions, algebraic manipulation, other purely abstract processes. If $a + b = x$ then $x = a - b$. If $ma/ca = IQ = 1.00$ then $Ma = CA$.

During your baby's first four months, did you have doubts that she really understood much that was happening around her? This parental reaction is not surprising. After all, although you knew when she was comfortable and uncomfortable, she probably showed few signs of actually thinking. Now, as her memory and attention span increase, you'll start to see evidence that she's not only absorbing information but also applying it to her day-to-day activities.

Discovering Cause and Effect

During this period, one of the most important concepts she'll refine is the principle of cause and effect. She'll probably stumble upon this notion by accident somewhere between 4 and 5 months. Perhaps while kicking her mattress, she'll notice the crib shaking. Or maybe she'll realize that her rattle makes a noise when she hits or waves it. Once she understands that she can cause these interesting reactions, she'll continue to experiment with other ways to make things happen.

Your baby will quickly discover that some things, like bells and keys, make interesting sounds when moved or shaken. When she bangs certain things on the table or drops them on the floor, she'll start a chain of responses from her audience, including funny faces, groans, and other reactions that may lead to the reappearance or disappearance of the object. Before long, she'll begin intentionally dropping things to see you pick them up. As annoying as this may be at times, it's one important way for her to learn about cause and effect and her personal ability to influence her environment.

It's important that you give your child the objects she needs for these experiments and encourage her to test her "theories." But make sure that everything you give her to play with is unbreakable, lightweight and large enough that she can't possibly swallow it. If you run out of the usual toys or she loses interest in them, plastic or wooden spoons, unbreakable cups, and jar or bowl lids and boxes are endlessly entertaining and inexpensive.

Discovering Object Permanence

Another major discovery that your baby will make during this period is that objects continue to exist when they're out of her sight, a principle called object permanence. During her first few months, she assumed that the world consisted only of things that she could see. When you left her room, she assumed you vanished; when you returned, you were a whole new person to her. In much the same way, when you hid a toy under a cloth or a box, she thought it was gone for good and wouldn't bother looking for it. But sometime after 4 months she'll begin to realize that the world is more permanent than she thought. You're the same person who greets her every morning. Her teddy bear on the floor is the same one that was in bed with her the night before. The block that you hid under the can did not actually vanish after all. By playing hiding games and observing the comings and goings of people and things around her, your baby will continue to learn about object permanence for many months to come.

Age 2 to 3 Years: Cognitive Development

Think back to your child's infancy and the early toddler months. That was a time when he learned about the world by touching, looking, manipulating and listening. Now the learning process has become more thoughtful. His grasp of language is increasing and he's beginning to form mental images for things, actions and concepts. He also can solve some problems in his head, performing mental trial-and-error instead of having to manipulate objects physically. And as his memory and intellectual abilities develop, he'll begin to understand simple time concepts, such as "You can play after you finish eating."

Your toddler also is starting to understand the relationship between objects. For instance, he'll be able to match similar shapes when you give him shape-sorting toys and simple jigsaw puzzles. He'll begin to recognize the purpose of numbers in counting objects, especially the number 2. And as his understanding of cause and effect develops, he'll become much more interested in winding up toys and turning lights and appliances on and off.

If we were to single out the major intellectual limitation at this age, it would be your child's feeling that everything that happens in his world is the result of something he has done. With a belief like this, it becomes very difficult for him to understand correctly such concepts as death, divorce or illness without feeling that he played some role in it. So if parents separate or a family member gets sick, children often feel responsible.

Reasoning with your 2-year-old is often difficult. After all, he views everything in extremely simple terms. He still often confuses fantasy with reality unless he's actively playing make-believe. For example, a wonderful story from Selma Fraiberg's The Magic Years talks about parents telling their 2-and-a-half-year-old that they would soon be flying to Europe. With a worried look on his face, the little boy said, "But my arms aren't strong enough to fly." Therefore, during this stage, be sure to choose your own words carefully. Comments that you think are funny or playful, such as "If you eat more ice cream, you'll explode," may actually panic him because he won't know you're joking.

3 to 4 Years- Cognitive Development

Your 3-year-old will spend most of her waking hours questioning everything that happens around her. She loves to ask, "Why do I have to...?" and she'll pay close attention to your answers as long as they're simple and to the point. Don't feel that you have to explain your rules fully; she can't yet understand such reasoning and isn't interested in it anyway. If you try to have this kind of "serious" conversation, you'll quickly see her stare into space or turn her attention to more entertaining matters, like a toy across the room or a truck passing outside the window. Instead, telling her to do something "because it's good for you" or "so you don't get hurt" will make more sense to her than a detailed explanation.

Answering Why Questions

Your child's more abstract "why" questions may be more difficult, partly because there may be hundreds of them each day, and also because some of them have no answers or none that you know. If the question is "Why does the sun shine?" or "Why can't the dog talk to me?" you can answer that you don't know, or invite her to look into the question further by finding a book about the sun or about dogs. Be sure to take these questions seriously. As you do, you help broaden your child's knowledge, feed her curiosity, and teach her to think more clearly.

When your 3-year-old is faced with specific learning challenges, you'll find her reasoning still rather one-sided. She can't yet see an issue from two angles, nor can she solve problems that require her to look at more than one factor at the same time. For example, if you take two equal cups of water and pour one into a short, fat container and the other into a tall, skinny one, she'll probably say the tall container holds more water than the short. Even if she sees the two equal cups to start with and watches you pour, she'll come up with the same answer. By her logic, the taller container is "bigger" and therefore must hold more. At around age 7, children finally understand that they have to look at multiple aspects of a problem before arriving at an answer.

Sense of Time

At about 3 years of age, your child's sense of time will become much clearer. She'll now know her own daily routine and will try hard to figure out the routines of others. For example, she may eagerly watch for the mail carrier who arrives every day but be perplexed that the garbage man comes only one day out of seven. She'll understand that certain special events, such as holidays and birthdays, occur every once in a while, but even if she can tell you how old she is, she'll have no real sense of the length of a year.

DEVELOPMENTAL EVALUATIONS

Although it's almost human nature to try to measure your child's intellectual progress, beware of over- or underestimating her reasoning or thinking ability without formal testing. It's easy to convince yourself that your friendly, happy, highly verbal 3-year-old is brilliant, while the quieter, more laid-back youngster playing next to her is not as bright. This may or may not be true, but the only way to know for sure is to have the child evaluated professionally. All pediatricians are familiar with these developmental evaluations, and some actually conduct formal assessments themselves. Many pediatricians prefer to refer children to testing specialists. So, if you believe your child is either gifted or lagging behind, ask your doctor about such testing. If your suspicions prove true, you may want to enter her in a special program geared to her individual needs and abilities.

Stimulating Brain Growth During the Third Year

Encourage playing with blocks and soft toys, which helps your child develop eye-hand coordination, fine-motor skills and a sense of competence.

Give consistent warm, physical contact - hugging, skin-to-skin, body-to-body contact to establish your toddler's sense of security and well-being.

Be attentive to your child's rhythms and moods. Respond to her when she is upset as well as when she is happy. Be encouraging and supportive, with firm discipline as appropriate, but without yelling or hitting. Provide consistent guidelines.

Talk or sing to your child during dressing, bathing, feeding, playing, walking and driving, using adult talk. Speak slowly and give your child time to respond. Try not to reply with "uh-huh" because your child will recognize when you're not listening; instead, expand upon your child's phrases.

— Be predictable; establish a pattern for mealtimes, naps and bedtime.

— Develop word associations by giving word labels to everyday objects and activities.

— Read to your child every day. Choose books that encourage touching and pointing to objects, and read rhymes, jingles and nursery stories.

— If you speak a foreign language, use it at home.

— Play calm and melodic music for your child.

— Listen to and answer your child's questions. Also, ask questions to stimulate decision-making processes.

— Begin to explain "safety" in simple terms. For example, feeling the heat from the stove teaches the meaning and danger of hot objects.

— Make sure other people who provide care and supervision for your baby understand the importance of forming a loving and comforting relationship with your child.

— Encourage your child to look at books and to draw.

— Help your child use words to describe emotions and to express feelings, such as happiness, joy, anger and fear.

— Spend time on the floor playing with your child every day.

— Choose quality child care that is affectionate, responsive, educational and safe. Visit your child-care provider frequently and share your ideas about positive caregiving.

Age 4 to 5 Years- Cognitive Development

By age 4, your child is beginning to explore many basic concepts that will be taught in greater detail in school. For example, she now understands that the day is divided into morning, afternoon and night, and that there are different seasons. By the time she enters kindergarten she may know some days of the week, and that each day is measured in hours and minutes. She also may comprehend the essential ideas of counting, the alphabet, size relationships (big versus small) and the names of geometric shapes.

There are many good children's books that illustrate these concepts, but don't feel compelled to rush things. There's no advantage to her learning them this early, and if she feels pressured to perform now, she may actually resist learning when she gets to school.

The best approach is to offer your child a wide rànge of learning opportunities. For instance, this is the perfect age to introduce her to zoos and museums, if you haven't already. Many museums have special sections designed for children, where she can actively experience the learning process.

Encouraging Interests

Respect her special interests and talents. If your child seems very artistic, take her to art museums and galleries or let her try a preschool art class. Also, if you know an artist, take her for a visit so she can see what a studio is like. On the other hand, if she's most interested in machines and dinosaurs, take her to the natural history museum, help her learn to build models, and provide her with construction kits that allow her to create her own machines. Whatever her interests, you can use books to help answer her questions and open her horizons even further. At this age, then, your child should be discovering the joy of learning so that she will be self-motivated when her formal education begins.

Answering Universal Questions

You'll also find that, in addition to exploring practical ideas, your 4-year-old probably will ask many "universal" questions about subjects, such as the origin of the world, death and dying, and the composition of the sun and the sky. Now, for example, is when you'll hear the classic question "Why is the sky blue?" Like so many other parents, you may have trouble answering these questions, particularly in simple Language your child will understand. As you grapple with these issues, don't make up answers but rely instead on children's books that deal with them. Your local library should be able to recommend age-appropriate books to help you.

Discover how you can support your child's cognitive development through play:

LEARNING, PLAY AND YOUR NEWBORN

Your newborn may not seem to be ready to learn and play, but all of his senses are working, and he's taking in the sensations, sounds, sights, and scents of his world. Since you comprise most of your newborn's world, teach him that it is a safe and loving one: fill his senses with lots of smiles, soothing sounds, and gentle caresses.

A parent's face is a newborn's first and most wonderful "toy." Your baby will love to fixate on your face, and it will be all the entertainment he needs for now. Provide plenty of smiles, and talk to him so he becomes familiar with your voice.

Your baby is curious about noises he hears, but none more so than your voice. He will learn very quickly to associate your voice with the things he needs: food, warmth, touch. He will listen carefully when you talk to him. If you react enthusiastically to the little noises he makes, he'll learn that you want him to express himself.

You will learn to recognize when your baby is alert and ready to learn and play and when he'd rather be left alone. Sometimes your baby will need to protect himself from overstimulation by "shutting down" for a bit.

Your newborn loves attention. Talk to him, read to him, stroke and cuddle him, and cover him with kisses to promote bonding and recognition. Even infants enjoy music and enjoy being read to. In fact, it seems that some babies actually remember songs and stories they heard repeatedly while still in the womb! But even if the song or story is not familiar to your baby, he will find it soothing.

Give your newborn simple toys that appeal to his sight, hearing, and touch. These include unbreakable crib mirrors, rattles, textured toys and musical toys. Choose toys and mobiles with contrasting colors and patterns; red and black patterns seem to appeal to newborns. A newborn's eyesight isn't very well developed, and strong contrasts will help him focus on his surroundings and keep his interest longer. Don't go overboard, though - offer your baby one plaything at a time. Too many can lead to confusion and overstimulation.

Play is not "play" to babies and children; play is how they learn about the world. Here are some ways to play with your newborn, but be sure to invent your own ways:

Dancing: Put on some music with a beat. Hold your baby's face close to your own, and gently sway and move to the tune. Feel free to sing softly.

Name game: Touch and name all the parts of your baby's body as he watches you. Make a game out of it by jumping from foot to foot or hand to hand.

Cuddling: Cradle your baby in your lap and gently stroke him in different rhythmic patterns.

If you have any questions about your newborn's ability to see or hear, you should bring them to your doctor's attention immediately. Even newborns can be tested using sophisticated equipment, if necessary. The sooner a potential problem is caught, the better it can be treated.

Learning, Play and Your 1-3 Month Old

Your 1-3 month old is more alert and aware of her surroundings than she was as a newborn. She already recognizes her parents' voices and faces, and she might be ready to respond to them with smiles. Your baby is ready to be an active participant in play. Your baby will carefully watch your facial expressions and listen to your voice. By listening to you, she is learning the importance of speech before she understands or repeats any words herself. She'll also learn during this period that she has the ability to vocalize, too; make sure to answer her coos and gurgles with your own sounds, and she will be more willing to express herself.

Now that your baby's hands are open, she'll begin to use them to learn about the world. She'll play with her fingers, bring her hands to her mouth, and try to swing at things within view. In this way, she is learning hand-eye coordination. When lying down, she'll stretch out her arms and legs - soon she will learn to grasp and kick! You will learn to recognize when your baby is alert and ready to learn and play and when she'd rather be left alone. Sometimes your baby will need to protect herself from overstimulation by "shutting down" for a bit.

Remember that play is not just "play" to babies and children. Play is how they learn, so be enthusiastic when your baby shows interest in playing. Take every opportunity to interact with her - provided she's in the mood. Don't overstimulate her with too many activities at once, and let her tell you when she's bored.

Your baby will enjoy listening to music, the sounds her toys make, and your singing - and she won't care whether or not you're any good. Her eyesight is improving, so she'll be fascinated by brightly colored pictures in books and the mobile above her crib. And she won't be able to take her eyes off herself if she has an unbreakable crib mirror.

As your baby's hands open, offer her a rattle to hold, and watch her search for the source of the rattle's sounds. Give her safe objects of different textures, shapes, sizes, colors, and weights to hold. Dangle objects above your baby and let her swat at them. To get her hands moving, clap them together. Move her legs like a bicycle with your hands. These body games will help her learn to control her movements.

LEARNING MILESTONES

Here are some milestones to look for during this period. By the end of three months, most babies:

— Smile at the sound of a parent's voice

— Smile at others

— Reach for, grasp, and hold objects

— Support their heads well

— Make babbling sounds

— Bring objects to their mouths

Babies develop at different rates, and there is a wide range of normal development. It is usually not a cause for concern when a baby misses a developmental milestone, but talk to your pediatrician if your baby seems behind.

Learning, Play, and Your 4-7 Month Old

Are you beginning to think that your 4-7 month old knows his name? You're probably right! His memory and attention span are increasing, and he's learning all sorts of things, even at this tender age.

Repetition is the name of the game for learning and play. Be prepared to play games, sing songs, and repeat actions over and over - you may be bored, but your baby is excited and stimulated!

Your baby is now beginning to understand the fundamentals of language. He's starting to pick out the components of your speech. He can hear and understand the different sounds you make and the way words form sentences. During this period, your baby will probably recognize and respond to his name - and it won't be long before he's learned to ignore the word "no"!

Babies generally begin to grasp the principle of cause-and-effect during this stage: he learns that his actions cause reactions and that he has the ability to make things happen. He will explore this concept by doing things - dropping an object, kicking his crib, yelling - over and over (and over) again. Although this may be frustrating, it is the way your baby is learning the effects his actions may have. He is not throwing his pacifier on the floor over and over in order to upset you. He is doing it in order to see if it appears before him again.

Now that your baby is beginning to realize that his actions have an effect on you, he will enjoy playing copycat games where you mimic what he says. He may also make the discovery that whenever he cries, he grabs your attention. This is mostly a good thing, but he will also use it on occasion when he is bored or frustrated. Rather than punish him for this, give him extra attention once he stops crying or fussing.

Your baby will learn a great deal about the properties of objects with his hands and mouth. He will struggle to reach objects and learn to transfer them from hand to hand. Every object he manages to reach will go in his mouth, so watch him carefully, and make sure the objects he can reach do not pose a choking hazard.

Reading to your baby gives him a chance to observe the way you talk and mimic the sounds you make. If the book has bright pictures, all the better. Be sure to show them to your baby and describe them. Books that will withstand some mouthing and chewing are best.

With your baby's more sophisticated motor skills, he is probably ready for toys that he can activate himself. Toys that make sounds when shaken are good, as are toys that teach cause-and-effect, such as pop-up toys. Puppets and other toys with finger holds will allow your baby to practice hand control. Be sure to read toy packages, and stick to the age guidelines suggested by the manufacturers.

Allow your baby to use his toys other than for their intended purpose. This stimulates his creativity and curiosity and encourages him to experiment.

Your baby is now beginning to understand the idea of object and person permanence, which means that objects and people continue to exist when they are out of sight. To demonstrate the principle of object permanence to your baby, try hiding a rattle, ball, or other toy under a towel or washcloth. See if he forgets all about the toy, or if he seems to grasp that the toy is simply hidden.

Captivate your baby with these simple games, and introduce your own:

Food Painting

Place some purified food on your baby's highchair tray, and let him "fingerpaint" with it. It's messy, but it's a lot of fun for your baby.

Peekaboo

This old standby will delight your baby, with his budding understanding of object and person permanence. Cover your face with your hands, then remove your hands and say: "Peekaboo, I see you!"

Hide-and-seek

Hide one of your baby's toys - but don't hide it very well - and encourage him to find it.

Sometimes your baby won't be in the mood to play, especially if he's tired, hungry, or cranky. The best thing to do is accept it and try again later.

By the end of this period, most babies will have learned to:

— Find an object that you have hidden

— Explore objects with their hands and mouths

Keep in mind, though, that there's a wide range of what's normal for babies. Some develop slower and some faster than others. If your child is a little behind what is considered normal, check with your pediatrician, but there's usually no cause for alarm.

Learning, Play and Your 8-12 Month Old

At this stage, you'll finally know that your baby was listening intently during all those "conversations" you had with her. She'll let you know that she understands what you say, and she'll respond with some "words" of her own. Keep those toys and games coming, and encourage your baby's exploration of the world. She is becoming more mobile and independent, but she won't want to be separated from you for very long.

Your baby will be demonstrating her understanding of what you say in several ways. She'll look at objects and people when you say their names, she'll crawl to toys that you ask her to find, and she'll do lots of pointing and gesturing in response to your words. She should already respond well to her own name, and she should look up when you firmly say, "NO!"

Your baby is continuing to explore the world of objects. Now that she can get around on her own, she'll find objects all over the house to pick up, shake, bang, throw and put in her mouth. Make sure the objects she finds are safe for all of these activities.

While this object exploration continues, she'll now understand the functions of some objects. She will recognize familiar objects and understand their purpose. For example, she might see a washcloth, know that it's for washing and may hold out her hands for it.

Your baby's new mobility is very exciting for her, but she's still learning the principles of object and person permanence, so she probably won't let you out of her sight for long. She'll play her own "peekaboo"-type games: she'll round the corner where she can no longer see you, then peek around the corner to make sure you're still there. Alternatively, she'll round the corner, then make a sound or do something that she knows will bring you to her.

She will know repetitive games like "pattycake" all the way through and will recognize the slightest variation. Her attention span will be short, but when you have her attention, she'll be able to point to a picture of a familiar object in a book when you say the object's name.

During this period, your baby begins to understand that people are unique individuals and separate from herself. This realization usually leads to separation anxiety; you may find that your baby has a difficult time being separated from you during the day and at bedtime. This is a perfectly normal development at this age and nothing to be concerned about - keep your exits short and sweet, and try to enjoy your baby's attachment to you while it lasts.

Your baby is getting around now, and that means she's finding all sorts of things to get into. Make sure your home is a safe place for learning and play by checking your childproofing efforts around the house. Be sure to supply your baby with toys and household objects that will develop her hand-eye coordination. As you probably know by now, just about anything is a toy to your baby. She is likely to play with egg cartons and empty cardboard boxes with the same enthusiasm she shows for large blocks, balls, stacking toys, and push-pull toys. Give her squeeze toys and cups and containers to splash around with during bath time.

Whenever you can get your busy baby to sit still for a moment, read to her from books with large, colorful illustrations. Encourage her to point out people and objects in the pictures.

Learning, Play and Your 1-2 Year Old

The majority of a child's social skills come as a result of play. During this stage, through play, your toddler will learn to form relationships and will begin to imitate adult actions and experiment with social activities. Play with your toddler as often as you can, and introduce him to a playmate or two.

Something that surprises many new parents is that babies do not learn steadily, but rather in spurts. They latch onto new ideas or skills quickly and practice them over and over, often seeming to forget much of what they've already learned. When they feel comfortable with the new skills, the old ones will be brought out and employed again. If your child enjoys an activity, don't hesitate to repeat it while also introducing him to new ones. Most babies say their first words by 12 months to 14 months, though some start sooner and some put it off for a while they focus on walking. Between 15 and 18 months, your baby will enjoy language games that ask him to identify things, such as: "Where's your ear?" and "Where is Mommy?"

Vocabulary will grow quickly, but pronunciation likely won't keep pace. Be patient and resist the temptation to correct your baby's pronunciation; instead, emphasize the correct pronunciation in your response. Soon he'll try to form sentences and questions with a combination of words and gestures. At this age, your baby will begin symbolic play for the first time - he'll imitate things he sees you do, and he'll use objects for their intended purpose. He'll attempt to brush his own hair with a hairbrush, or he'll use the phone to "talk" to you.

Your child may fluctuate between wanting independence and wanting to cling to you. This is absolutely normal - children this age are learning about their separateness from you but are still very dependent on you for comfort and reassurance. Give your child as much freedom as you safely can, and be there when he needs to be a "baby" for a while.

Remember that safety is your responsibility. Your baby may be learning how things work and what things do, but that doesn't mean he understands what is harmful or dangerous. Take a look at your house from you toddler's point of view, and periodically check your childproofing efforts.

"Lessons" should be in the form of games to keep your child's interest, and they should be stopped at the first sign of boredom. Removing the pressure from a lesson or playtime will teach your baby that learning is fun, and he'll look forward to the activity in the future.

Your toddler will be very interested in experimenting with social activities and imitating adult actions in make-believe play. Give him toys that encourage symbolic play, such as kitchen sets, dolls, and dress-up clothes. Allow him to "shadow" you as you do chores or everyday activities, and give him some "chores" to do, such as brushing his teeth or putting on his shoes.

Great toys for exploring and experimenting include trains, play garden tools, outdoor toys (swing sets, slides), blocks, large crayons, baskets, and pots and pans. Shape sorters, pegboards, nesting toys and simple puzzles allow your baby to enhance his analytical skills. As always, read to your child. He will now follow along with the story and point out objects and people in the picture.

Your toddler will be fascinated with other children at this age. One or two playmates at a time is plenty. Don't expect your toddler to "play" with other children in a cooperative way or to be enthusiastic about sharing toys; he's not mature enough yet. You'll need to be there at all times to diffuse conflict. Nevertheless, children this age can learn a lot from each other by imitation.

Enthusiasm and applause can be two very powerful tools for stimulating your child to learn. Pay attention to his successes, no matter how small, and he'll be encouraged to try for more.

Although babies develop at different rates, most babies this age:

- Engage in symbolic play
- Speak at least 15 words by 18 months and use two-word sentences by age 2
- Imitate your expressions and actions and follow your instructions
- Attempt more independent activity
- Check with your pediatrician if you are concerned that your baby has not reached any of the above milestones

— More information on intelligence and learning:

MULTIPLE INTELLIGENCES

In Howard Gardner's Frames of Mind, he proposes that there are seven main areas in which all people have special skills; he calls them intelligences. His research at Harvard University was in response to the work that Alfred Binet had done in France around 1900. Binet's work led to the formation of an intelligence test; we are all familiar with the "intelligence quotient," or "IQ," the way that intelligence is measured on his test.

Both of these tests look predominantly at two types of intelligences: verbal and math. If a person does well on these, s/he is considered "intelligent," and is a candidate for one of the better colleges or universities. But what about everyone else? How many of you who are reading these words have used the phrase "not good at taking tests," when talking either about yourself or your child?

The Multiple Intelligences (MI) theory proposes that there are other measures of intelligence beside these two. This information to you so that you can understand that while many teachers have some knowledge of MI theory, most of our schools are not fully set up to use it to the advantage of all students. That being the case, perhaps you can either (1) be involved in helping your child's teachers and school to provide a more balanced program that develops his intelligences that are not more included in the curriculum or (2) find activities outside of the school environment in which your child can develop his dominant areas of intelligence.

You should also know that MI theory posits that each of us has, to some degree or another, all of these intelligences. Some of them are simply more developed than others. Furthermore, we are all able to improve our ability in each of these areas.

Howard Gardner stresses that the intelligences are equal in their importance. In alphabetical order, they are:

Bodily-kinesthetic: using one's body to solve problems and express ideas and feelings. Actors, athletes, and dancers use their whole bodies in this way, much the same way that craftspeople, sculptors, and mechanics use their hands.

These questions can determine if an adult has a strength in Bodily-Kinesthetic Intelligence:

— Do you regularly participate in a sport or some physical activity?

— Is it difficult to sit still for long periods of time?

— Do you enjoy working with your hands in creating things?

— Do you find that ideas and solutions to problems come to you while you are exercising or doing some sort of physical activity?

— Do you enjoy spending your free time outdoors?

— Do you speak with your hands or other body gestures?

— Do you learn more about things by touching them?

— Do you enjoy thrilling amusement park rides such as the roller coaster and other activities like this?

— Do you think of yourself as being well-coordinated?

— In order to learn a new skill, do you have to practice it to learn it, rather than read about it or see it in a video?

These are some questions to determine if children may be exhibiting a well-developing Bodily-Kinesthetic Intelligence. Does your child:

— excel in more than one sport?

— move various body parts when required to sit still for long periods of time?

— have the ability to mimic others' body movements?

— enjoy taking things apart and putting them back together?

— have a hard time keeping hands off objects?

— enjoy running, jumping, or other physical activities?

— show skill in activities that require fine-motor coordination, such as origami, making paper airplanes, building models, finger-painting, clay, or knitting?

— use his body well to express himself?

Interpersonal: perceiving the moods, feelings, and needs of others. It includes salespeople, teachers, counselors, and those we have come to call the helping professions. These questions can determine if an adult has a strength in Interpersonal Intelligence:

— Have people always come to you for advice?

— Have you always preferred group sports to solo sports?

— Do you usually prefer talking to other people about a problem, rather than figure it out on your own?

— Do you have at least three close friends?

— Do you prefer social activities over individual pursuits?

— Do you enjoy teaching others what you can do well?

— Are you considered to be a leader, either by yourself or others?

— Do you feel comfortable in a crowd?

— Do you prefer to spend your time with others than alone?

These are some questions to determine if children may be exhibiting a well-developing Interpersonal Intelligence. Does your child:

— enjoy socializing with friends?

— seem to be a natural leader?

— empathize easily with others, which leads to his give advice to friends who come to him with problems?

— seem to be street-smart?

— enjoy belonging to organizations?

— enjoy teaching other kids - either peers or younger ones?

— have two or more close friends?

— serve as a magnet for social activities with others?

Intrapersonal: turning inward with a well-developed self-knowledge and using it successfully to navigate oneself through the world.

These questions can determine if an adult has a strength in Intrapersonal Intelligence:

— Do you regularly spend time alone meditating, reflecting, or thinking about important life questions?

— Have you attended counseling sessions or personal growth seminars to learn more about yourself?

— Do you have a hobby or interest that you keep to yourself?

— Have you set goals for yourself regularly?

— Do you have a realistic view of your strengths and weaknesses?

— Would you prefer spending time by yourself rather than with many people around you?

— Do you keep a diary or journal to record the events of your inner life?

— Are you either self-employed or have you given serious consideration to starting your own business?

These are some questions to determine if children may be exhibiting a well-developing Intrapersonal Intelligence. Does your child:

— show a sense or independence or a strong will?

— have a realistic sense of her abilities and weaknesses?

— do well when left alone to play or study?

— "march to the beat of a different drummer" in living and learning?

— have a hobby or interest she doesn't talk about much?

— have a good sense of self-direction?

— prefer working alone to working with others?

— accurately express how he is feeling?

— learn from failures and successes?

— have good self-esteem?

Linguistic: using words, either orally or written, in an effective manner. This intelligence is associated with storytellers, politicians, comedians, and writers.

These questions can determine if an adult has a strength in Linguistic Intelligence:

— Have you always enjoyed books and given them importance?

— Do you hear words in your head before you speak or write them?

— Do you enjoy talk shows more than television or movies?

— Do you enjoy word games, puns, rhymes, tongue-twisters, and poetry?

— Do you have a highly developed vocabulary and enjoy knowing words that other people do not know?

— In your own education, did you enjoy subjects related to words and ideas, such as English and social studies, more than math and science?

— Have you enjoyed learning to read or speak other languages?

— In your speech, do you refer to information that you have read or heard about?

— Have you been praised, recognized, or paid for your writing?

These are some questions to determine if children may be exhibiting a well-developing Linguistic Intelligence. Does your child:

- — write better than average for her age?
- — enjoy telling stories and jokes?
- — have a good memory for names, places, dates, and other information?
- — enjoy word games, either visually or auditorally?
- — enjoy reading books?
- — spell better than other children the same age?
- — appreciate rhymes, puns, tongue twisters?
- — enjoy books on tape without needing to see the book itself?
- — enjoy hearing stories without seeing the book?
- — have an excellent vocabulary for his age?
- — communicate thoughts, feelings, and ideas well?

Logical-Mathematical: understanding and using numbers effectively, as well as having good powers to reason well. Exemplars are mathematicians, scientists, computer programmers, and accountants. These questions can determine if an adult has a strength in Logical-Mathematical Intelligence:

- — Have you always done math in your head easily?
- — When you were in school, were math and/or science your best subjects?
- — Do you enjoy playing games that require logical thinking?
- — Do you set up experiments to see "what if" in your course of jobs around the house or at work?
- — Do you look for logical sequences and patterns, with the belief that almost everything has a logical explanation?
- — Do you read science periodicals or keep track of the latest scientific developments?

— Do you like finding logical flaws in things that people say and do?

— Do you feel the need to have things measured, categorized, analyzed, or quantified in some way?

These are some questions to determine if children may be exhibiting a well-developing Logical-Mathematical Intelligence. Does your child:

— demonstrate curiosity about how things work?

— have fun with numbers?

— enjoy math at school?

— enjoy math and/or computer games?

— play and enjoy strategy games such as chess and checkers, brain teasers, or logic puzzles?

— easily put things into categories?

— like to do experiments, either at school when assigned or on her own?

— show an interest in visiting natural history or discovery-type museums and exhibits?

Musical: relating in a wide range of ways to music. This can take many forms, as a performer, composer, critic, and music-lover. These questions can determine if an adult has a highly developed Musical Intelligence:

— Do you have a pleasant singing voice?

— Can you tell when a musician plays a note off-key?

— Do you frequently listen to music?

— Do you play a musical instrument?

— Was it easy for you to learn to play a musical instrument?

— Do you think your life would not be as rewarding without music?

— Do you usually have music going through your mind?
— Can you keep time to music?
— Do you know the tunes to many different songs or musical selections?

These are some questions to determine if children may be exhibiting a well-developing Musical Intelligence. Does your child:

— tell you when she recognizes that music is off-key?
— easily remember song melodies and sing them?
— have a pleasant singing voice, either alone or in a chorus?
— play a musical instrument?
— speak or move in a rhythmical way?
— hum or whistle to himself?
— tap on the tabletop or desktop while working?
— show sensitivity to noises in the environment?
— respond emotionally to music she hears?

Naturalist Intelligence: excellent at recognizing and classifying both the animal and plant kingdoms, as well as showing understanding of natural phenomena.

These questions can determine if an adult has a strength in Naturalist Intelligence:

— Do you like to spend time in nature?
— Do you belong to a volunteer group related to nature?
— Do you enjoy having animals around the house?
— Are you involved in a hobby that involves nature, such as bird watching?
— Can you easily tell the differences among species of flora and fauna?

— Do you read books or magazines, or watch television shows or movies that feature nature?

— On vacation, do you prefer natural settings to cultural attractions?

— Do you enjoy visiting zoos, aquariums, or other places where the natural world is studied?

— Do you enjoy working in your garden?

These are some questions to determine if children may be exhibiting a well-developing Naturalist Intelligence. Does your child:

— talk about favorite pets or preferred natural spots?

— enjoy nature preserves, the zoo, or natural history museum?

— show sensitivity to natural formations? (Note that in urban environments, this type of "formation" can include cultural icons.)

— like to play in water?

— hang around the pet in school or at home?

— enjoy studying environment, nature, plants, and animals?

— speak out about animal rights and earth preservation?

— collect bugs, flowers, leaves, or other natural things to show to others?

Spatial: perceiving the visual-spatial world in an accurate way, so as to be able to work in it effectively. The people who do this cover a wide range of fields that, upon first glance, do not seem to have much in common. Compare, for example, hunters, sailors, engineers, inventors, and surgeons to interior decorators, architects, painters, and sculptors.

LANGUAGE AND LITERACY DEVELOPMENT

Language is a beautiful gift. With it we can share our wants, our needs, our thoughts, our feelings, and everything that makes us human. If you spend time with a child, you have the power to give and nurture this gift of communication. Many factors affect the rate at which a child develops language. Sometimes language development slows down while a child is learning other skills , such as standing or walking. In other words, the bulk of the child's concentration and energy may be going to gross motor development at this point with little reserve for the development of language.

The amount and kind of language the child hears may also affect the rate of language development. For example, if the child is hearing two languages at home , his or her brain is trying to learn two sets of vocabulary, process two sets of speech sounds, and understand two sets of grammatical rules. That is a lot of work! It may take longer to begin talking, and still the child may at first feel comfortable speaking in only one of the languages. Some children who are immersed in a new language at school may be silent for a long period of time.

The rate of language development may also be affected by how people respond to the child . For example, the child whose communication attempts are greeted with eye contact, acknowledgement ("Uh huh. Tell me more. What else happened?"), and expansion of his or her ideas is likely to develop language faster than the child whose communication attempts receive little or no response.

As children develop language, they typically go through the same stages of development. These stages are presented below. The exact age at which a specific child goes through a certain stage varies because of the factors given above, but the order of the stages is pretty much the same from child to child.

Talk and read to your child a lot. By reading books to your child as early as six weeks you can also put him/her on the road to reading.

Birth to One Year

A baby's first attempts at communicating emotions and needs

are through crying. Parents quickly learn how to differentiate hunger cries from those indicating tiredness or a wet diaper.

By three months , the baby will turn his or her head towards voices and recognize parents' voices (e.g., by stopping crying when parents are speaking to him or her). Expressively, the baby indicates contentment and/or amusement by smiling. He or she repeats sounds (e.g., cooing).

At 4 to 6 months , the baby notices new sounds such as the vacuum and telephone. He or she responds to "no" and to changes in tone of voice. He or she pays attention to music. Early sound discrimination skills are beginning to emerge. Sounds have a more speech-like babble to them (e.g., babamama). When playing alone or with parents, the child makes gurgling sounds. He or she tells you by sound or gesture when he or she wants something.

At 7 months to one year , the infant is beginning to recognize his or her name (e.g., turning the head or looking up when you say his or her name). The child listens when spoken to. The baby is beginning to recognize common words (e.g., "cup," "juice," and "bottle"), and to respond to requests like "Come here." Vocabulary and concepts needed for reading begin here. Expressively, the infant imitates speech sounds, and he or she may have 1 or 2 words (e.g., "bye-bye," "baba" for bottle, and "no") by one year. The baby more frequently uses speech or non-crying sounds to get and keep attention. Babbling has long and short groups of sounds such as "mama maaa."

What You Can Do to Help

Always check your child's ability to hear, and pay attention to ear problems and infections , especially when they keep occuring. Even at this early age, there is much that you can do to encourage language development. Reinforce your baby's communication attempts by looking at him or her, speaking, and imitating his or her vocalizations. Imitate his or her laughter and facial expressions. Teach your baby to imitate actions, such as peekaboo, clapping, blowing kisses, pat-a-cake, itsybitsy spider, and waving "bye-bye." These games teach turn taking that is needed for conversation.

Talk while you are doing things, such as dressing, bathing, and feeding. "Mommy is washing Sam's hair." "Sam is eating carrots." "Oh, these carrots are good!" Talk about where you are going, what you will do once you get there, and who and what you'll see. "Sam is going to grandma's house. Grandma has a dog. Sam will pet the dog." Talk about colors (e.g., "Sam's hat is red"). Practice counting. Count toes and fingers. Count steps as you go up and down them. Teach animal sounds.

One to Two Years

Your child is now able to participate more actively in listening to simple stories (e.g., Brown Bear Brown Bear, by Bill Martin Jr.), songs (e.g., "Twinkle, Twinkle Little Star" and "If You're Happy and You Know It"), and rhymes ("Humpty Dumpty"). He or she can follow simple commands (e.g., "Roll the ball" and "Blow a kiss"). Your child can point to a few body parts. He or she can point to pictures of things in a book when you name them.

Your child's vocabulary is increasing, and he or she says more and more words every month. Some 1-to-2-word questions, such as "What's that?" , are used. Your child is beginning to put 2 words together; for example, "more juice" and "get up." He or she uses many different consonant sounds at the beginning of words.

What You Can Do to Help

Continue to talk while doing things and going places. When taking a walk in the stroller, for example, point to familiar objects (e.g., cars, trees, and birds) and say their names. "I see a dog. The dog says 'woof.' This is a big dog. This dog is brown."

Use simple but grammatical speech that is easy for your child to imitate.

Take a sound walk around your house or in the baby's room. Introduce him/her to Timmy Clock, who says "t-t-t-t." Listen to the clock as it ticks. Find Mad Kitty Cat who bites her lif and says "f-f-f-f" or Vinnie Airplane who bites his lip, turns his voice motor on and says "v-v-v-v." These sounds will be old friends when your child is introduced to phonics in preschool and kindergarten.

Make bath time "sound play time" as well. You are eye-level with your child. Play with Peter Tugboat, who says, "p-p-p-p." Let her feel the air of sounds as you make them. Blow bubbles and make the sound "b, b, b, b." Feel the motor in your throat on this sound. Engines on toys can make a wonderful "rrr-rrr-rrr" sound.

Expand on single word utterances. For example, if your child says "car," you respond by saying, "You're right! That is a big red car." Continue to find time to read to your child every day. Try to find books with large pictures and 1-2 words or a simple phrase or sentence on each page. When reading to your child, take time to name and describe the pictures on each page. Have your child point to pictures that you name. Ask your child to name pictures. He or she may not respond to your naming requests at first. Just name the pictures for him or her. One day, he or she will surprise you by coming out with the picture' s name.

Two to Three Years

Your child is beginning to understand differences in meaning. For example, he or she understands the difference between "up and down," "in and out," and "big and little."Your child notices noises, such as the doorbell ringing, the telephone, and sounds on the television. He or she follows requests or directions that have two parts to them. Expressively, your child has a word for almost everything, and he or she is beginning to use

2-3 word phrases to talk about things and ask questions. He or she directs attention to or asks for objects by naming them. Familiar listeners understand your child's speech most of the time.

Three to Four Years

Your child now talks in sentences of 4 or more words. He or she talks about activities (e.g., what was done in preschool that day) easily and fluently, without repeating syllables or words. Now people outside of the family are able to understand him or her. Your child understands and answers simple who, what, and where questions. He or she hears you when you call from another room. He or she can hear the television or radio at the same loudness level as other family members.

Encouraging continued language development gets really fun at this age. Cut out pictures from old catalogs. Then make silly pictures by gluing parts of different pictures together in an improbable way. For example, glue a picture of a dog to the inside of a car as if the dog is driving. After a good laugh, help your child explain what is silly about the picture. You can do the same thing with family photos that you do not need anymore.

Continue to sort pictures and items into categories, but increase the challenge by asking your child to point out the item that does not belong in a category. For example, a baby does not belong with a dog, cat and mouse. Tell your child that you agree with his or her answer because a baby is not an animal.

Continue to expand vocabulary and the length of your child' s utterances by reading, singing, talking about what you are doing and where you are going, and saying rhymes. Now you can read books that have a simple plot, and you can talk about the story line with your child. Help your child to retell the story or act it out with props and dress-up clothes. Tell him or her your favorite part of the story and ask for his or her favorite part.

Continue to look at family pictures, and now have your child explain what is happening in each one. Work on comprehension skills by asking your child questions. Have him or her try to fool you with his or her own questions. Make this game playful by pretending that you have been fooled by some of his or her really hard questions.

Expand on social communication and story-telling skills by "acting out" typical scenarios (e.g., cooking food, going to sleep, and going to the doctor) with a dollhouse and its props. Do the same type of role-playing activity when playing dress-up. As always, ask your child to repeat what he or she has said if you do not understand it completely. This shows that what he or she says is important to you.

Four to Five Years

Your child now pays attention to short stories and answers simple questions about them. He or she understands most of what is said at home and at school. Your child communicates easily with other children and adults. Sentences give details, and use adult-like grammar. When explaining something or telling a story, your child sticks to the topic and strings together ideas in an understandable sequence. Your child may still have some errors in pronunciation (e.g., /s/, /r/, /l/, /v/, /ch/, /sh/, /th/ , /j/, /z/), but is still easy to understand.

What You Can Do to Help

Always give full attention to your child when he or she is speaking, and acknowledge, praise, and encourage him or her after he or she has spoken. Before you speak to your child, be sure to get his or her undivided attention. Pause after speaking, allowing him or her to respond to what you have said. As always, continue to build on your child' s vocabulary. Provide definitions for new words, and use them in context. "This vehicle is riding on the highway. It is a car. A bus is another kind of vehicle. So are a train and an airplane."

Encourage your child to ask for an explanation if he or she does not understand what a word means. Talk about spatial relationships (e.g., between, under, first in line) and encourage your child to do the same. Point out things that are the same or different. Play games that incorporate these concepts which he/she will meet later in the classroom in reading readiness.

Continue sorting items into categories. Now try to sort them by pointing out more subtle differences between objects (e.g., rocks that are smooth versus those that are rough, heavy vs. light, big vs. small). Again, have your child identify the object that does not belong in a given category, but now ask him or her to explain why the item does not belong. Expand on social communication and narration skills (telling a story) by role-playing. Play house, doctor, and store using dialogue, props, and dress-up clothes. Do the same with a dollhouse and its props, acting out scenarios and making the dolls talk.

Continue to read stories with easy-to-follow plots. Help your child predict what will happen next in the story. Act out the stories, and put on puppet shows of the stories. Have your child draw a picture of a scene from the story, or of a favorite part. You can do the same thing with videos and television shows, as these also have plots. Ask "wh" questions (who, what, when, where, or why) and monitor his/her response. Expand on your child' s comprehension and expressive language skills by playing "I Spy." "I spy something round on the wall that you use to tell the time." After your child guesses what you have described, have him or her give you clues about something that he or she sees.

Give your child 2-step directions (e.g., "Get your coat from the closet and put it on"). Encourage your child to give directions to explain how he or she has done something. For example, ask your child to explain how he or she made a structure out of Lego blocks. When playing doctor, ask you child to explain what he or she did to give the baby a check-up. Draw a picture, and write down your child's story as he tells it. He will soon grasp the power of storytelling and written language.

Play age-appropriate board games with your child (e.g., "Candyland" or "Chutes and Ladders").

Have your child help you plan and discuss daily activities. For example, have him or her make a shopping list for the grocery store, or help you plan his or her birthday party. Ask his or her opinion. What do you think your cousin would like for his birthday? What kind of fruit do we need to buy at the store?

Oral Language Components

Oral language, the complex system that relates sounds to meanings, is made up of three components: the phonological, semantic, and syntactic. The phonological component involves the rules for combining sounds. Speakers of English, for example, know that an English word can end, but not begin, with an -ng sound. We are not aware of our knowledge of these rules, but our ability to understand and pronounce English words demonstrates that we do know a vast number of rules.

The semantic component is made up of morphemes, the smallest units of meaning that may be combined with each other to make up words (for example, paper + s are the two morphemes that make up papers), and sentences. A dictionary contains the semantic component of a language, but also what words (and meanings) are important to the speakers of the language.

The syntactic component consists of the rules that enable us to combine morphemes into sentences. As soon as a child uses two morphemes together, as in "more cracker," she is using a syntactic rule about how morphemes are combined to convey meaning. Like the rules making up the other components, syntactic rules become increasingly complex as the child develops. From combining two morphemes, the child goes on to combine words with suffixes or inflections (-s or -ing, as in papers and eating) and eventually creates questions, statements, commands, etc. She also learns to combine two ideas into one complex sentence, as in "I'll share my crackers if you share your juice."

Of course speakers of a language constantly use these three components of language together, usually in social situations. Some language experts would add a fourth component: pragmatics, which deals with rules of language use. Pragmatic rules are part of our communicative competence, our ability to speak appropriately in different situations, for example, in a conversational way at home and in a more formal way at a job interview. Young children need to learn the ways of speaking in the day care center or school where, for example, teachers often ask rhetorical questions. Learning pragmatic rules is as important as learning the rules of the other components of language, since people are perceived and judged based on both what they say and when they say it.

Teachers' Role in Language Development

Parents and caregivers need to remember that language in the great majority of individuals develops very efficiently. Adults should try not to focus on "problems," such as the inability to pronounce words as adults do (for example, when children pronounce r's like w's). Most children naturally outgrow such things, which are a tiny segment of the child's total repertoire of language. However, if a child appears not to hear what others say to her; if family members and those closest to her find her difficult to understand; or if she is noticeably different in her communicative abilities from those in her age range, adults may want to seek advice from specialists in children's speech, language and hearing.

— Teachers can help sustain natural language development by providing environments full of language development opportunities. Here are some general guidelines for teachers, parents, and other caregivers:

— Understand that every child's language or dialect is worthy of respect as a valid system for communication. It reflects the identities, values, and experiences of the child's family and community.

— Treat children as if they are conversationalists, even if they are not yet talking. Children learn very early about how conversations work (taking turns, looking attentively, using facial experiences with conversing adults.

— Encourage interaction among children. Peer learning is an important part of language development, especially in mixed-age groups. Activities involving a wide range of materials should promote talk. There should be a balance between individual activities and those that nurture collaboration and discussion, such as dramatic play, block-building, book-sharing, or carpentry.

— Remember that parents, caregivers, teachers, and guardians are the chief resources in language development. Children learn much from each other, but adults are the main conversationalists, questioners, listeners, responders, and sustainers of language development and growth in the child-care center or classroom.

— Continue to encourage interaction as children come to understand written language. Children in the primary grades can keep developing oral abilities and skills by consulting with each other, raising questions, and providing information in varied situations. Every area of the curriculum is enhanced through language, so that classrooms full of active learners are hardly ever silent.

Participating in Language Development

Your participation in your child's language development will become even more important after 6 or 7 months, when he begins actively imitating the sounds of speech. Up to that point, he might repeat one sound for a whole day or even days at a stretch before trying another. But now he'll become much more responsive to the sounds he hears you make, and he'll try to follow your lead. So introduce him to simple syllables and words like "baby," "cat," "dog," "go," "hot," "cold" and "walk," as well as "Mama" and "Dada." Although it may be as much as a year before you can interpret any of his babbling, your baby can understand many of your words well before his first birthday.

A baby with a partial hearing loss still can be startled by loud noises or will turn his head in their direction, and he may even respond to your voice. But he will have difficulty imitating speech. If your child does not babble or produce a variety of sounds, alert your pediatrician. If he has had frequent ear infections he might have some fluid remaining in his inner ear, and this could interfere with his hearing.

A very young baby's hearing can be checked by using special equipment, but your observations are the early warning system that tells whether such testing is needed. If you suspect a problem, you might ask your pediatrician for a referral to a children's hearing specialist.

Ages 1 to 2 Years: Language Development

Your baby learns language in stages. From birth he receives information about language by hearing people make sounds and watching how they communicate with one another. At first he is most interested in the pitch and level of your voice. When you talk to him in a soothing way, he'll stop crying because he hears that you want to comfort him. By contrast, if you shout out in anger he probably will cry, because your voice is telling him something is wrong. By 4 months, he'll begin noticing not only the way you talk but the individual sounds you make. He'll listen to the vowels and consonants, and begin to notice the way these combine into syllables, words and sentences.

As well as receiving sounds, your baby also has been producing them from the very beginning, first in the form of cries and then as coos. At about 4 months, he'll start to babble, using many of the rhythms and characteristics of his native language. Although it may sound like gibberish, if you listen closely you'll hear him raise and drop his voice as if he were making a statement or asking a question. Encourage him by talking to him throughout the day. When he says a recognizable syllable, repeat it back to him and then say some simple words that contain that sound. For example, if his sound of the day is "bah," introduce him to "bottle," "box," "bonnet" and "Baa, Baa, Black Sheep."

Your participation in your child's language development will become even more important after 6 or 7 months, when he begins actively imitating the sounds of speech. Up to that point, he might repeat one sound for a whole day or even days at a stretch before trying another. But now he'll become much more responsive to the sounds he hears you make, and he'll try to follow your lead. So introduce him to simple syllables and words like "baby," "cat," "dog," "go," "hot," "cold" and "walk," as well as "Mama" and "Dada." Although it may be as much as a year before you can interpret any of his babbling, your baby can understand many of your words well before his first birthday.

When to Call the Pediatrician for Language Development

If he doesn't babble or imitate any sounds by his seventh month, it could mean a problem with his hearing or speech development. A baby with a partial hearing loss still can be startled by loud noises or will turn his head in their direction, and he may even respond to your voice. But he will have difficulty imitating speech. If your child does not babble or produce a variety of sounds, alert your pediatrician. If he has had frequent ear infections he might have some fluid remaining in his inner ear, and this could interfere with his hearing.

A very young baby's hearing can be checked by using special equipment, but your observations are the early warning system that tells whether such testing is needed. If you suspect a problem, you might ask your pediatrician for a referral to a children's hearing specialist.

Age 2 to 3 Years- Language Development

Your 2-year-old not only understands most of what you say to her but also speaks with a rapidly growing vocabulary of 50 or more words. Over the course of this year, she'll graduate from two- or three-word sentences ("Drink juice," "Mommy want cookie") to those with four, five or even six words ("Where's the ball, Daddy?" or "Dolly sit in my lap"). She's also beginning to use pronouns (I, you, me, we, they) and understands the concept of "mine" ("I want my cup," "I see my mommy"). Pay attention to how she is using language to describe ideas and information, and to express her physical or emotional needs and desires.

It's human nature to measure your toddler's verbal abilities against those of other children her age, but try to avoid this. There's more variation at this time in language development than in any other area. While some preschoolers develop language skills at a steady rate, others seem to master words in an uneven manner. And some children are naturally more talkative than others. This doesn't mean that the more verbal children are necessarily smarter or more advanced than the quieter ones, nor does it mean that they have richer vocabularies. In fact, the quiet child may know just as many words but be choosier about speaking them. As a general rule, boys start talking later than girls, but this variation, like most others mentioned above, tends to even out as children reach school age.

Without any formal instruction, just by listening and practicing, your child will master many of the basic rules of grammar by the time she enters school. You can help enrich her vocabulary and language skills by making reading a part of your everyday routine. At this age, she can follow a story line and will understand and remember many ideas and pieces of information presented in books. Even so, because she may have a hard time sitting still for too long, the books you read to her should be short. To keep her attention, choose activity-oriented books that encourage her to touch, point and name objects or to repeat certain phrases. Toward the end of this year, as her language skills become more advanced, she'll also have fun with poems, puns or jokes that play with language by repeating funny sounds or using nonsense phrases.

For some youngsters, however, the language-development process does not run smoothly. In fact, about one in every 10 to 15 children has trouble with language comprehension and/or speech. For some, the problem is caused by hearing difficulty, low intelligence or lack of verbal stimulation at home. In most cases, though, the cause is unknown. If your pediatrician suspects your child has difficulty with language, she'll conduct a thorough physical exam and hearing test and, if necessary, refer you to a speech/language or early-childhood specialist for further evaluation. Early detection and identification of language delay or hearing impairment is critically important so that treatment can begin before the problem interferes with learning in other areas. Without identifying the difficulty and doing something about it, the child may have continuous trouble with classroom learning.

3 to 4 Years- Language Development

At 3, your child should have an active vocabulary of 300 or more words. He'll be able to talk in sentences of five or six words and imitate most adult speech sounds. At times, he'll seem to be chattering constantly, a phenomenon that may sometimes disturb you but which is essential to his learning of new words and gaining experience in using and thinking with them. Language allows him to express his thoughts, and the more advanced he is in speaking and understanding words, the more tools he'll have for thinking, creating and telling you about it.

At this age, your child's speech should be clear enough that even strangers can understand most of what he says. Even so, he still may mispronounce as many as half the speech sounds he uses. For example, he may use w for r ("wabbit," "wice," "wose"), d for th ("dis," "dat," "den"), or t for any sounds he has trouble with ("tee" for three, "tik" for six). The sounds b, p, m, w and h will only begin to emerge midway through this year, and it may take months after that for him to perfect his use of them.

Expanding Vocabulary

You should be able to see how your child uses language to help him understand and participate in the things going on around him. For instance, he can name most familiar objects, and he'll freely ask "What's this?" when he can't call something by name. You can help him expand his vocabulary by providing additional words that he might not even request. For example, if he points to a car and says, "Big car," you might answer, "Yes, that's a big gray car. Look how shiny the surface is." Or if he's helping you pick flowers, describe each one he collects. "That's a beautiful white-and-yellow daisy, and that's a pink geranium."

You also can help him use words to describe things and ideas he can't see. When he's describing the "monster" in his dream, for example, ask him if the monster is angry or friendly. Ask him about the monster's color, where he lives, whether he has friends. Not only will this help your child use words to express his thoughts, but also it may help him overcome his fear of such strange and frightening images.

Explaining Pronouns

Your 3-year-old is still learning to use pronouns, such as "I," "me," "mine" and "you." As simple as these words seem, they're difficult ideas to grasp, because they indicate where his body, possessions or authority ends and someone else's begins. And to complicate matters, the terms change depending on who's talking. Often, he may use his name instead of saying "I" or "me." Or when talking to you, he may say "Mommy" instead of "you." If you try to correct him (for example, by suggesting, "Say 'I would like a cookie' "), you'll only confuse him more, because he'll think you're talking about yourself. Instead, use these pronouns correctly in your own speech. Say "I would like you to come" instead of "Mommy would like you to come." Not only will this help him learn the correct use of these words, but it will help him establish a sense of you as an individual apart from your role as Mommy.

Age 4 to 5 Years- Language Development

At about age 4, your child's Language skills will blossom. She'll now be able to pronounce most of the sounds in the English Language, with the following exceptions: f, v, s and z probably will remain difficult for her until midway through age 5, and she may not fully master sh, l, th and r until age 6 or later.

Your preschooler's vocabulary will have expanded to around 1,500 words by now, and it will grow by another 1,000 or so over the course of this year. She can now tell elaborate stories using relatively complex sentences of up to eight words. And she will tell you not only about things that happen to her and things she wants but also about her dreams and fantasies.

Handling Bossiness and Swear Words

Your child will probably pick up many swear words at this age. From her point of view, these are the most powerful words of all. She hears adults say them when they are most angry or emotional, and whenever she uses them herself, she gets quite a reaction. The best way to stop this behavior is to be a good role model and make a conscious effort not to use these words, even when you are stressed. In addition, try to minimize your child's using these words without drawing too much attention to them. She probably has no idea what these words really mean; it's just the energy of them that she enjoys.

Handling Insults

When your child's upset, you may find she'll use words as insults. Of course this is certainly preferable to physical violence, although it can be quite disturbing to you. Remember, though, that when your child uses these words, she's disturbed too. If she says "I hate you!" what she really means is "I am very angry, and I want you to help me sort out my feelings." By getting angry and shouting back at her you'll only make her feel more hurt and confused. Instead, remain calm and tell her you know she doesn't really hate you. Then let her know that it's okay to feel angry, and talk about the events leading up to her outburst. Try giving her the words that will allow her to tell you how she feels.

If the insults she chooses are mild ones, the best response may be a joke. For example, let's presume she calls you a "wicked witch," you might laugh and respond, "And I'm just boiling up a pot of bat's wings and frog's eyes. Care to join me for supper?" This kind of humor is an excellent way to take the edge off her anger as well as your own.

Of course, sometimes your preschooler doesn't have to say anything offensive to try your patience, her constant chatter can do it just as quickly. One solution at these moments is to redirect her verbal energy. For instance, instead of allowing her to chant mindless sound rhymes, teach her some limericks or songs, or take time out to read some poems. This will help her learn to pay more attention to the words she speaks and will boost her appreciation for the written Language as well.

COMMUNICATION AND YOUR NEWBORN

Do you remember your baby's very first cry? From the moment

of birth, your baby began to communicate. At first, your newborn's cries may seem like a foreign language to you. But before you know it, you will learn your baby's "language" and be able to provide him with what he needs. Right after birth, you introduced your baby to your own ways of communicating: touching, soothing, holding and making facial gestures. Your baby will learn your "language," just as you will learn his.

Your baby is born with the ability to cry, which is how he will do most of his communicating for a while. Your baby's cries generally tell you that something is wrong in his world: his belly is empty, his bottom is wet, his feet are cold, he's tired, he needs to be held and cuddled. Soon you will be able to recognize which need your baby is expressing and respond accordingly. In fact, sometimes what a baby needs can be identified by his cry - for example, the "I'm hungry" cry may be short and low-pitched, while "I'm upset" may sound choppy.

Your baby may also cry when he is overwhelmed by all of the sights and sounds of the world. Sometimes he may cry for no apparent reason at all. Don't be too upset when your baby cries and you aren't able to console him immediately: crying is one of his ways of shutting out stimuli when he's overloaded. Crying is your baby's main method of communication, but he is also capable of other, more subtle forms. Learning to recognize them is exciting and rewarding and can strengthen your bond with your baby.

A newborn can differentiate between the sound of a human voice and other sounds. Try to pay attention to how he responds to your voice. He already associates your voice with care: food, warmth, touch. If he's crying in his bassinet, see how quickly your approaching voice quiets him. See how closely he listens when you are talking to him in loving tones. He may not yet coordinate looking and listening, but even if he stares into the distance, he'll be paying close attention to your voice as you speak. He may subtly adjust his body position or facial expression, or even move his arms and legs in time with your speech. Sometime during your newborn's first month, you may get a glimpse of his first smile and perhaps hear his first laugh or giggle. What welcome additions these are to his communication repertoire!

As soon as you hold your baby after birth, you will begin to communicate with each other by exchanging your first glances, sounds, and touches. Your newborn is already learning about the world through his senses. As the days after birth pass, your newborn will become accustomed to seeing you and will begin to focus on your face. His senses of touch and hearing are especially important, though. Your infant will be curious about noises he hears, but none more so than the spoken voice. Talk to your baby whenever you have the chance. Even though he doesn't understand what you're saying, your calm, reassuring voice is what he needs to feel safe. With almost every touch your newborn is learning about life, so provide him with lots of tender kisses, and he'll find the world a soothing place to be.

Communicating with a newborn is really a matter of meeting his needs. Always respond to your newborn's cries - he cannot be spoiled with too much attention. Your prompt response when your baby communicates will let him know that he's important and worthy of attention. There will probably be times when you have met all of your baby's needs, yet he continues to cry. Don't despair - your baby may be overly stimulated, have gastric distress, or may have too much energy and need a good cry. It is common for babies to have a fussy period at the same time every night, generally between early evening and midnight. This can be very upsetting, but the good news is that it's short-lived; most babies outgrow it around three months.

There are some things you can try to soothe your baby. Some babies are comforted by motion, such as rocking or being walked back and forth across the room, while others respond to sounds, like soft music or the hum of a vacuum cleaner. It may take some time to find out what best comforts your baby during these stressful periods.

Communication and Your 1-2 Year Old

Language development really takes off during this time, especially as your baby approaches his second birthday. He is better able to comprehend what you say and express what he wants. He will take joy in his ability to understand more complex directions - and he won't hesitate to give you directions.

Most babies say their first words toward the beginning of this period, though some start even sooner and others don't start talking until they are nearly 2 years old. If your baby is preoccupied with learning to walk, he may push talking to the back burner; this is not unusual and nothing to be alarmed about.

Your baby may have learned fragments of dozens of words that probably won't be recognizable yet. When he gets around to talking, though, he'll probably progress quickly. He'll soon be able to point at something familiar and say its name, and recognize names of familiar people, objects and body parts. By two years, he may use phrases and even two- to four-word sentences.

No matter when your child says his first words, it's a sure bet he'll be understanding much of what you say to him well before that. He should be able to respond to commands and should be fully aware of the names of familiar objects and family members. You will undoubtedly find yourself struggling with your toddler to do as you say, only to have him ignore you or scream in protest. He's merely testing your limits and his degree of control. By 18 months, he will probably have mastered saying "no" with authority, and by age 2 he may throw a tantrum when he's unwilling to do something you ask. He'll also show signs of possessiveness, and you'll frequently hear "mine" or see tears if something is taken away or you show attention toward someone else.

Parents who suspect that their child is having trouble with hearing, language acquisition, or speech clarity should not hesitate to talk with their child's doctor. A hearing test can be one of the first referral/diagnostic measures to determine if there might be a hearing problem. If you suspect that your child has a hearing problem at any age, she should undergo an evaluation. Two years of age is not too young for a referral for a speech/language evaluation, particularly if a child is not following directions or answering "yes" or "no" to simple questions.

After an evaluation, the speech-language pathologist (an expert who evaluates and treats speech and language disorders) may recommend direct therapy, referral to a developmental pediatrician if there is suspicion of a global developmental delay (delay in more than one area of development, including gross motor, fine motor, problem solving, language, and social skills), early intervention services, or simply a follow-up assessment to see if a child will catch up.

Communication and Your 4-5 Year Old

Communicating with your child, from infancy onward, is one of the most pleasurable and rewarding experiences for both parent and child. Children are avid learners at all ages, absorbing information through daily interactions and experiences with other children, adults and the world. Between the ages of 4 and 5, many children begin to enter preschool or kindergarten programs, making language competency necessary for better learning in the classroom.

The more interactive conversation and play a child is involved in, the more a child learns. Reading books, singing, playing word games, and simply talking to your child will increase his vocabulary while providing increased listening opportunities. Here are a few suggestions to help improve your child's communication skills:

Help your child relate to books by selecting stories that mirror families like yours or people from your cultural or ethnic group. Make a special spot for books, magazines and other reading material where your child is able to reach books without help.

Children ages 4 to 5 are also able to understand that letters and numbers are symbols of real things and ideas, "write" as a way to tell stories and offer information, and "read" on their own. Sentence structures now incorporate up to eight words, while vocabulary is between 1,000 and 2,000 words. Ninety percent of the time, speech should be clearly intelligible; although there may be some developmental sound errors and stuttering, particularly among boys.

Young preschoolers make comments and requests and tell others what to do. By 4 or 5, they should know the names and gender of family members and other personal information. They often play with words and make up silly words and stories. Parents worried about a child's hearing, language acquisition, or speech clarity should not hesitate to consult with their child's doctor. A hearing test may be one of the first referral/diagnostic measures to determine if there has been a hearing problem. If the doctor suspects a specific communication deficit or delay, a referral for a speech-language evaluation may be recommended. If a child appears to be delayed in all areas of development, it may be more appropriate to refer him to a developmental pediatrician and/or psychologist.

After an evaluation, the speech-language pathologist (an expert who evaluates and treats speech and language disorders) may recommend direct therapy, preschool special education services, or possible referrals to an audiologist (a hearing specialist), developmental pediatrician, and/or psychologist.

Communication problems among children in this age group include:

— Hearing difficulties

— Problems following complex/compound directions

— Difficulty with conversational interaction

— Poor vocabulary acquisition

— Difficulty learning preschool concepts, such as colors and counting

— Stuttering

— Difficulties with grammar and syntax

— Unclear speech

Common problems, such as stuttering, often may be a developmental process that some children will outgrow. For others, more intensive therapy may be needed. These communication problems can be helped by medical professionals, such as speech pathologists, therapists, or your child's doctor.

CHILDREN AND BILINGUALISM

According to the 1990 United States Census, one in seven, or 31.8 million, people speak a language other than English in their home. In the past, second generation children were encouraged to adopt the customs, culture, and language of what was identified as the majority culture. Today, with greater recognition and celebration of cultural differences, people are more likely to maintain and share their primary language with their children and to promote bilingualism as a reflection of ethnic pride and identity.

Different people use the term in different ways. For some, bilingualism means an equal ability to communicate in two languages. For others, bilingualism means the ability to communicate in two languages, but with the possibility of greater skills in one language. There are different theories on the "best" way to teach a child to use two languages. Most researchers agree that a child who is exposed simultaneously to two languages at an early age will naturally learn to use both languages. For example, if you are interested in children's materials in Spanish, go to www.csusm.edu/csb/.

Children can be expected to go through some periods of mixing the two languages and borrowing vocabulary to express ideas, sometimes within the same sentence. This occurs because vocabulary may exist in one language but not in the other. Or, words from one language may convey a message that is not easily translated into the other language. A separation of the two languages will occur gradually.

Children may also experiment with the two languages to create special effects or to express themselves in specific settings. For example, one language may be identified as less formal and used for information about events related to home and family. The other language may be identified as more formal and used for activities outside the home. There may also be periods when one language is used more than the other. The opposite occurs with a change in the environment.

Children may not be equally skilled in both languages. It is common for there to be greater understanding than actual use of one language. Less confusion will occur if children learn to associate the two languages differently, for example, if one language is used while speaking to the mother and the other while speaking to the father.

In general, speech-language problems are less likely to occur when both languages are introduced early and simultaneously. There is a greater possibility of problems if children are introduced to a second language during the preschool years after another language was used exclusively. Some people believe that, if a second language is fully developed, the development of the first language may be slowed or even regress. Others believe that the skill level of the second language will develop only to that of the first.

Parents who are concerned about their child's speech-language development should contact a speech-language pathologist certified by the American Speech-Language-Hearing Association. An appropriate evaluation of skills will include evaluation of both languages, and will be completed by a bilingual speech-language pathologist or one who has knowledge of the rules and structure of both languages and the assistance of a translator or interpreter. The discovery that a child has some delays in both languages does nor necessarily indicate a need for professional services. These delays may be characteristic of a two-language learner.

Children with Communication Disorders

The term communication disorders encompasses a wide variety of problems in language, speech, and hearing. Speech and language impairments include articulation problems, voice disorders, fluency problems (such as stuttering), aphasia (difficulty in using words, usually as a result of a brain injury), and delays in speech and/or language. Speech and language delays may be due to many factors, including environmental factors or hearing loss.

Hearing impairments include partial hearing and deafness. Deafness may be defined as a loss sufficient to make auditory communication difficult or impossible without amplification. There are four types of hearing loss. Conductive hearing losses are caused by diseases or obstructions in the outer or middle ear and can usually be helped with a hearing aid. Sensorineural losses result from damage to the sensory hair cells of the inner ear or the nerves that supply it and may not respond to the use of a hearing aid. Mixed hearing losses are those in which the problem occurs both in the outer or middle ear and in the inner ear. A central hearing loss results from damage to the nerves or brain.

Many communication disorders result from other conditions such as learning disabilities, cerebral palsy, mental retardation, or cleft lip or cleft palate.

Characteristics of Children with Communication Disorders

A child with speech or language delays may present a variety of characteristics including the inability to follow directions, slow and incomprehensible speech, and pronounced difficulties in syntax and articulation. Syntax refers to the order of words in a sentence, and articulation refers to the manner in which sounds are formed. Articulation disorders are characterized by the substitution of one sound for another or the omission or distortion of certain sounds. Stuttering or dysfluency is a disorder of speech flow that most often appears between the ages of 3 and 4 years and may progress from a sporadic to a chronic problem. Stuttering may spontaneously disappear by early adolescence, but speech and language therapy should be considered.

EDUCATIONAL IMPLICATIONS OF COMMUNICATION DISORDERS

Many speech problems are developmental rather than physiological, and as such they respond to remedial instruction. Language experiences are central to a young child's development. In the past, children with communication disorders were routinely removed from the regular class for individual speech and language therapy. This is still the case in severe instances, but the trend is toward keeping the child in the mainstream as much as possible. In order to accomplish this goal, teamwork among the teacher, speech and language therapist, audiologist, and parents is essential. Speech improvement and correction are blended into the regular classroom curriculum and the child's natural environment.

Amplification may be extremely valuable for the child with a hearing impairment. Students whose hearing is not completely restored by hearing aids or other means of amplification have unique communication needs. Children who are deaf are not automatically exposed to the enormous amounts of language stimulation experienced by hearing children in their early years.

For deaf children, early, consistent, and conscious use of visible communication modes such as sign language, finger spelling, and cued speech and/or amplification and aural/oral training can help reduce this language delay.

Some educators advocate a strict oral approach in which the child is required to use as much speech as possible, while others favor the use of sign language and finger spelling combined with speech, an approach known as total communication. There is increasing consensus that whatever system works best for the individual should be used. Many children with hearing impairments can be served in the regular classroom with support services. In addition to amplification, instructional aids such as captioned films and high interest/low vocabulary reading materials are helpful. For most children with hearing impairments, language acquisition and development are significantly delayed, sometimes leading to an erroneously low estimate of intelligence.

Students whose physical problems are so severe that they interfere with or completely inhibit communication can frequently take advantage of technological advances that allow the individual to make his or her needs and wants known, perhaps for the first time.

4

Classroom Interventions for Children's Learning Disabilities

Children with learning disabilities exhibit a wide range of symptoms. These include problems with reading, mathematics, comprehension, writing, spoken language, or reasoning abilities. Hyperactivity, inattention and perceptual coordination may also be associated with learning disabilities but are not learning disabilities themselves.

The primary characteristic of a learning disability is a significant difference between a child's achievement in some areas and his or her overall intelligence.

— The learning disabled have difficulties with academic achievement and progress. Discrepancies exist between a person's potential for learning and what he actually learns.

— The learning disabled show an uneven pattern of development (language development, physical development, academic development and/or perceptual development).

— Learning problems are not due to environmental disadvantage.

— Learning problems are not due to mental retardation or emotional disturbance.

CAUSES LEARNING DISABILITIES

Little is currently known about the causes of learning disabilities. However, some general observations can be made:

— Some children develop and mature at a slower rate than others in the same age group. As a result, they may not be able to do the expected school work. This kind of learning disability is called "maturational lag."

— Some children with normal vision and hearing may misinterpret everyday sights and sounds because of some unexplained disorder of the nervous system.

— Injuries before birth or in early childhood probably account for some later learning problems.

— Children born prematurely and children who had medical problems soon after birth sometimes have learning disabilities.

— Learning disabilities tend to run in families, so some learning disabilities may be inherited.

— Learning disabilities are more common in boys than girls, possibly because boys tend to mature more slowly.

— Some learning disabilities appear to be linked to the irregular spelling, pronunciation, and structure of the English language. The incidence of learning disabilities is lower in Spanish or Italian speaking countries.

Learning disasaffect five general areas:

Spoken language: delays, disorders, and deviations in listening and speaking.

Written language: difficulties with reading, writing and spelling.

Arithmetic: difficulty in performing arithmetic operations or in understanding basic concepts.

Reasoning: difficulty in organizing and integrating thoughts.

Memory: difficulty in remembering information and instructions.

Among the symptoms commonly related to learning disabilities are:

— poor performance on group tests

— difficulty discriminating size, shape, color

— difficulty with temporal (time) concepts

— distorted concept of body image

— reversals in writing and reading

— general awkwardness

— poor visual-motor coordination

— hyperactivity

— difficulty copying accurately from a model

— slowness in completing work

— poor organizational skills

— easily confused by instructions

— difficulty with abstract reasoning and/or problem solving

— disorganized thinking

— often obsesses on one topic or idea

— poor short-term or long-term memory

— impulsive behavior; lack of reflective thought prior to action

— low tolerance for frustration

— excessive movement during sleep

— poor peer relationships

— overly excitable during group play

— poor social judgment

— inappropriate, unselective, and often excessive display of affection
— lags in developmental milestones (e.g. motor, language)
— behavior often inappropriate for situation
— failure to see consequences for his actions
— overly gullible; easily led by peers
— excessive variation in mood and responsiveness
— poor adjustment to environmental changes
— overly distractible; difficulty concentrating
— difficulty making decisions
— lack of hand preference or mixed dominance
— difficulty with tasks requiring sequencing

How learning disability affect the parents of the child?

Research indicates that parental reaction to the diagnosis of learning disability is more pronounced than in any other area of exceptionality. Consider: if a child is severely retarded or physically handicapped, the parent becomes aware of the problem in the first few weeks of the child's life. However, the pre-school development of the learning disabled child is often uneventful and the parent does not suspect that a problem exists. When informed of the problem by elementary school personnel, a parent's first reaction is generally to deny the existence of a disability. This denial is, of course, unproductive. The father tends to remain in this stage for a prolonged period because he is not exposed to the child's day-to-day frustrations and failures.

A parent may move from stage-to-stage in random. Some parents skip over stages while others remain in one stage for an extended period. These stages are as follows:

Denial: "There is really nothing wrong!" "That's the way I was as a child—not to worry!" "He'll grow out of it!"

Blame: "You baby him!" "You expect too much of him." "It's not from my side of the family."

Fear: "Maybe they're not telling me the real problem!" "Is it worse than they say?" "Will he ever marry? go to college? graduate?"

Envy: "Why can't he be like his sister or his cousins?"

Mouring: "He could have been such a success, if not for the learning disability!"

Bargaining: "Wait 'till next year!" "Maybe the problem will improve if we move! (or he goes to camp, etc.)."

Amger: "The teachers don't know anything." "I hate this neighborhood, this school...this teacher."

Guilt: "My mother was right; I should have used cloth diapers when he was a baby." "I shouldn't have worked during his first year." "I am being punished for something and my child is suffering as a result."

Isolation: "Nobody else knows or cares about my child." "You and I against the world. No one else understands."

Flight: "Let's try this new therapy—Donahue says it works!" "We are going to go from clinic to clinic until somebody tells me what I want to hear.!"

Again, the pattern of these reactions is totally unpredictable. This situation is worsened by the fact that frequently the mother and father may be involved in different and conflicting stages at the same time (e.g., blame vs. denial; anger vs. guilt). This can make communication very difficult. The good news is that with proper help, most LD children can make excellent progress.

RIGHTS AND RESPONSIBILITIES OF PARENTS OF CHILDREN WITH DISABILITIES

Parents of children with disabilities have a vital role to play in the education of their children. This fact is guaranteed in federal legislation that specifies the right of parents to participate in the educational decision-making process. As your child progresses through educational systems, knowing and following through on your rights and responsibilities ensures that you are a contributing partner with professionals who will influence your child's future. This brochure provides you with an introduction to your rights and responsibilities in the special education process.

What Are Your Rights in the Special Education Process?

Public Law 101-476 (IDEA) clearly defines the rights of children with disabilities and their parents. A basic provision of the law is the right of parents to participate in the educational decision-making process. Your rights, more specifically, include the following:

— Your child is entitled to a free, appropriate public education meaning it is at no cost to you as parents, and it meets the unique educational needs of your child.

— You will be notified whenever the school wishes to evaluate your child, wants to change your child's educational placement, or refuses your request for an evaluation or a change in placement.

— You may request an evaluation if you think your child needs special education or related services.

— You should be asked by your school to provide parent consent — meaning you understand and agree in writing to the evaluation and initial special education placement for your child. Your consent is voluntary and may be withdrawn at any time.

— You may obtain an independent evaluation if you disagree with the outcome of the school's evaluation.

— You may request a reevaluation if you suspect your child's current educational placement is no longer appropriate. The school must reevaluate your child at least every 3 years, but your child's educational program must be reviewed at least once during each calendar year.

— You may have your child tested in the language he or she knows best. For example, if your child's primary language is Spanish, he or she must be tested in Spanish. Also, students who are hearing impaired have the right to an interpreter during the testing.

— The school must communicate with you in your primary language. The school is required to take whatever action is necessary to ensure that you understand its oral and written communication, including arranging for an interpreter if you are hearing impaired or if your primary language is not English.

— You may review all of your child's records and obtain copies of these records, but the school may charge you a reasonable fee for making copies. Only you, as parents, and those persons directly involved in the education of your child will be given access to personal records. If you feel any of the information in your child's records is inaccurate, misleading, or violates the privacy or other rights of your child, you may request that the information be changed. If the school refuses your request, you then have the right to request a hearing to challenge the questionable information in your child's records.

— You must be fully informed by the school of all the rights provided to you and your child under the law.

— You may participate in the development of your child's Individualized Education Program (IEP) or, in the case of a child under school age, the development of an Individualized Family Service Plan (IFSP). The IEP and

IFSP are written statements of the educational program designed to meet your child's unique needs. The school must make every possible effort to notify you of the IEP or IFSP meeting and arrange it at a time and place agreeable to you. As an important member of the team, you may attend the IEP or IFSP meeting and share your ideas about your child's special needs, the type of program appropriate to meeting those needs, and the related services the school will provide to help your child benefit from his or her educational program.

— You may have your child educated in the least restrictive school setting possible. Every effort should be made to develop an educational program that will provide the greatest amount of contact with children who are not disabled.

— You may request a due process hearing to resolve differences with the school that could not be resolved informally.

— What Are Your Responsibilities in the Special Education Process?

— Parental responsibilities to ensure that a child's rights are being protected are less clearly defined than are parental rights. These responsibilities vary depending on the child's disabling condition and other factors. Some of the following suggestions may be helpful:

— Develop a partnership with the school or agency. You are now an important member of the team. Share relevant information about your child's education and development. Your observations and suggestions can be a valuable resource to aid your child's progress.

— Learn as much as you can about your rights and the rights of your child. Ask the school to explain these rights as well as the regulations in effect for your district and state before you agree to a special education program for your child. Contact disability organizations for their publications on special education rights.

— Ask for clarification of any aspect of the program that is unclear to you. Educational and medical terms can be confusing, so do not hesitate to ask.

— Make sure you understand the program specified on the IEP or IFSP before agreeing to it or signing it. Ask yourself if what is planned corresponds with your knowledge of your child's needs.

— Consider how your child might be included in the regular school activities program. Do not forget areas such as lunch, recess, art, music, and physical education.

— Monitor your child's progress. If your child is not progressing, discuss it with the teacher and determine whether the program should be modified. As a parent, you can initiate review of your child's educational program.

— Discuss with the school or agency any problems that may occur with your child's assessment, placement, or educational program. It is best to try to resolve problems directly with the agency or school. In some situations, you may be uncertain as to how you should resolve a problem. All states have advocacy agencies that can usually provide you with the guidance you need to pursue your case.

— Keep records. There may be many questions and comments about your child that you will want to discuss, as well as meetings and phone conversations you will want to remember. It is easy to forget information useful to your child's development and education if it is not written down.

— Join a parent organization. In addition to the opportunity to share knowledge and support, a parent group can be an effective force on behalf of your child. Many times parents find that as a group they have the power to bring about needed changes that strengthen special services.

Many organizations have information to help guide you through the special education process. Since the specific criteria and procedures used by school districts vary, it is important to familiarize yourself with the information provided by state and local agencies. You will find your local school district's director of special education and his or her staff helpful in accessing such information and guiding you through the process

Preparing Children with Disabilities for School

The transition from preschool to school can be difficult for a child with disabilities. The preschool environment characterized by small groups and individual attention is replaced by classrooms with more children, fewer adults per child, and greater demand for adapting to general classroom procedures and working independently. Parents and teachers from both receiving and sending programs need to be involved in placement as well as scheduling and facilitating the move. P.L. 99-457 recognizes the importance of preparing children and their families by requiring that specific steps be addressed in each child's individualized family service plan (IFSP) for children from birth through age 2 or individualized education program (IEP) for preschool children.

Role of Family members

Family members play a key role in providing information about the child's abilities, strengths and weaknesses, and interests. Parental insights complement information obtained from preschool sources and provide a broader picture of the child's capabilities and needs. Identifying specific ways for parents to be involved in the process is essential to a good transition.

Parents may act as teachers, partners, decisionmakers, and/or advocates. They are teachers when they reinforce the skills acquired in preschool, partners when they communicate needs with school personnel, and decisionmakers when they participate in the IEP process. Parents can help prepare the child for the transition to public school by maintaining and generalizing skills necessary for the transition. They also serve as a bridge between the two programs, visiting the new program with their child, helping the child to become familiar with the new setting, and discussing concerns and fears connected with the upcoming change. They can also help bridge the gap by arranging visits with former preschool friends and teachers as well as with new classmates. Parents can help their child develop skills in following directions, playing independently, attending to task, and self-care. These skills will help prepare the child for the new setting.

Role of Teachers

Sending and receiving teachers both play important roles in the transition process. Teacher attitudes, instructional priorities, and communication with parents and other members of the transition team will determine the quality of the child's transition. Sending and receiving teachers may have different goals and priorities, but they play complementary roles in preparing the child for the move from preschool to the general school setting.

The sending teacher should find out what skills the child will need in order to function adequately in the new setting and implement a program for preparing the child to develop those skills. Familiarity with the receiving program is essential in order to design an appropriate transition curriculum. The sending teacher can gain a better understanding of prerequisite skills by visiting the receiving classroom. For children placed in an integrated setting, behavioral requirements for successful functioning have been assessed and are referred to as survival skills. Teaching survival skills as part of the preschool curriculum helps prepare the child for the demands of the general school setting. Skill-building activities should be developmentally appropriate for each child.

The success of the transition preparation is ultimately determined by the child's adaptation to the new environment. The receiving teacher's attitude toward and experience with children with disabilities may be factors in the success of the child's placement. Some flexibility will probably be required on the teacher's part in order to adjust expectations and adapt to the child's special needs. The sending and receiving teachers will have the continuing role of acting as liaisons between programs and with parents. Good communication and clearly defined goals will facilitate the preparation for the child's move from preschool to the general school setting.

Elements of a Successful Transition Process

The Capstone Transition Process is one model that provides clear guidelines for the transition process. The first activity initiates long-range planning by establishing a "transition timeline." This timeline serves as a guide for accomplishing transition activities and can be set up in chart form to track activities. The Capstone Transition Process addresses specific activities beginning 12 months before the move to a new program. The process includes preparation, implementation, and evaluation activities. The initial steps of the process are designed to prepare the participants for their role in the transition. Steps include notifying and preparing parents and teachers from both the sending and receiving programs. Data on the child's needs are collected or updated.

A profile of communication procedures, available services, prerequisite skills, and teacher expectations is developed from existing information. The preparation phase of the process culminates with the development by the transition team of an IEP for use as the basis of educational programming in the new setting. Following the IEP meeting, the timeline provides reminders for the transfer of information and records to the receiving program. The final step calls for evaluation of the effectiveness of the process.

IMPROVING LEARNING ABILITY

This section provides information related to assisting parents in helping their children and teenagers to improve their learning and study skills. Techniques on how to develop motivation, attention, comprehension, organization and creativity are provided. Information and materials to bring basic skills in reading, math and written language up to or above grade level are also available.

Attention Deficit Disorder

Some children have more trouble paying attention in class and completing academic assignments than others. It is estimated that from 3 to 10 percent of the population has a condition known as Attention Deficit Disorder (ADD) or Attention Deficit Hyperactivity Disorder (ADHD). This disorder is said to be found present more often in boys than girls (3:1). Not every student having trouble completing assignments or who is squirmy and disruptive to the class has Attention Deficit Disorder. This is particularly true in the lower grades where many of the students have not reached a level of maturity sufficient to meet the demands of the classroom.

Some studies have found that a first grade teacher may rate as many as 50% of the boys in the class as having Attention Deficit Disorder using common clinical questionnaires. Thus, educators may mislabel behavior as being abnormal when it actually may be within the low end of normal development. Actually, normal attention span seems to develop in three stages.

First, the child's attention is said to be overly exclusive. This is a term used by psychologists to describe attention that is focused on a single object for a long period of time while tuning out all other stimuli. An example of this would be a baby who focuses totally on a button or a pin on the clothing of the person holding him. A child whose attention becomes stuck at this stage of development might be diagnosed as having autism.

Second, a child's attention develops to where it is overly inclusive. This refers to a very wide span of attention that is constantly and rapidly changing from one object to another such as a toddler who is running from one toy to the next never able to stay with any one toy for any period of time. A child who is stuck at the second stage of attention span development might be diagnosed as having attention deficit disorder.

Third, the child develops selective attention where he is able to shift focus at will from being inclusive to being very exclusive. This is a mature pattern of attention and concentration that is required to be successful in a classroom learning environment.

There are a variety of causes for poor attention, concentration and impulse control. A partial list would include the following:

— Immaturity and slow psychological development

— Learning disabilities such as dyslexia

— Anxiety

— Depression

— Low thyroid

— Low motivation

— Lack of sufficient sleep

— Poor nutrition

— Boredom due to lack of challenge

If your child has had the symptoms of poor attention, concentration and/or poor impulse control for some time, you should have your child examined by his physician. In addition to a physical exam and appropriate lab tests, the physician may have the parents and teachers complete questionnaires about the child's behavior. He may also refer you to a school psychologist or a mental health professional for further assessment. Depending on the outcome of the entire evaluation, one or several specific medical, educational or psychological recommendations may be given.

Causes

The symptoms of Attention Deficit Disorder are caused by a neurological dysfunction within the brain. Several studies using PET scans have confirmed that there is a definite difference in brain functioning between a group of individuals diagnosed with Attention Deficit Disorder and those without it. The underlying physiological mechanism which causes Attention Deficit Disorder is still not thoroughly understood and remains under scientific study. It is presumed that brain chemistry is out of balance and that specific chemicals called neurotransmitters may be lacking in individuals with Attention Deficit Disorder.

In two research studies, when the Attention Deficit Disorder subjects were given their medication and rescanned, their brain activity appeared to be much more like that of the normal group. Another interesting finding of these studies is that the areas of the brain in the Attention Deficit Disorder group where brain activity was lower than in the normal group are known to be associated with such functions as attention and concentration as well as planning and organization. These are the very functions that are impaired in persons with Attention Deficit Disorder. PET scans are considered to be very invasive because they involve the injection of radioactive material into the individual and therefore are not appropriate for diagnostic purposes. To date only two studies of Attention Deficit Disorder using PET scans have been conducted.

Attention Deficit Disorder may be either inherited or acquired. Recent research in genetics has definitely shown that Attention Deficit Disorder runs in families. The specific chromosomes that are involved have not been fully identified though more clues are being discovered as research progresses. Like many other disorders, a child may have the symptoms while his parents may not. Usually there are aunts, uncles, grandparents or cousins with the disorder.

Attention Deficit Disorder may be acquired through various conditions that cause insult (damage) to the brain. During pregnancy and delivery these include the use of drugs during pregnancy, toxemia, infectious diseases, overexposure to radiation, prematurity, complicated delivery. After birth these include meningitis, encephalitis, seizures from fever, head injury and lead toxicity. Excessive use of sweets does not cause Attention Deficit Disorder though it may make the symptoms worse for some individuals who already have the condition

Since many children with Attention Deficit Disorder appear to "outgrow" the condition it can also be looked at as a developmental disorder. Some children seem to develop the ability to pay attention and concentrate later than others just as there is a wide range to timing for developing the ability to walk, talk or be potty trained. In the past, children with Attention Deficit Disorder and other learning disabilities were labeled as "underachievers" in their younger years and then reclassified as "late bloomers" when their development in this area finally caught up with their peers.

Today, many adults look back and realize they had a learning disability or Attention Deficit Disorder as a child which they may or may not have been able to overcome. Some people with Attention Deficit Disorder become very successful. However, others continue to have significant symptoms as adults. Some were not able to adapt or compensate on their own and did not receive any help from teachers, parents or professionals. These individuals often dropped out of school and may today be social drop outs of one kind or another. It is for this last reason that children who exhibit signs of Attention Deficit Disorder or learning disabilities should receive all the help necessary and available to spare them the frustration and anguish often associated with these conditions.

Treatments

Medication

Medications such as stimulants have long been employed in

the treatment of Attention Deficit Disorder. These medications are presumed to improve a chemical imbalance in the brain which is causing the symptoms. The PET scan studies do show that the brain functioning of Attention Deficit Disorder patients does improve and appears to be more like the normal group after they have taken their prescribed medication.

Medications usually used to treat Attention Deficit Disorder stimulate the production of two neurotransmitters known as dopamine and norephinephrine. Specific neurotransmitters (brain chemicals) are necessary to carry a nerve impulse (message) along a neuropathway (circuit). When a neurotransmitter is under supplied, a message may be stopped short of its intended destination. When this happens, the function regulated by that circuit may not work as well as it should.

Brain circuits, like those of a computer are either on or off. When some circuits are on they make something happen such as helping a child focus on a learning situation. When other circuits are on they prevent something from happening. For instance, some circuits prevent emotional reactions to situations. If the circuit is not on or only partially on, the child may react too quickly to a minor incident which may lead to a temper tantrum.

Medications that treat Attention Deficit Disorder are not tranquilizers or sedatives. They do not slow down the nervous system. They actually stimulate various areas of the brain to be more active so that the attention and concentration functions and the self-control functions work better. The use of stimulant medications helps to keep the circuits on when they should be turned on.

Medication is the most frequently employed treatment method for Attention Deficit Disorder. It is often employed along with psychological techniques such as behavior modification and patient/family education. Focus is a psychoeducational program that designed so that it can be used as either an adjunct or as an alternative to medication.

Diet and Nutrition

Scientific research does not support the use of diet and nutrition in the treatment of Attention Deficit Disorder. At one time, the Feingold Diet was very popular and seen as an alternative to medication. The elimination of sweets may help some individuals to reduce the symptoms but is usually not sufficient to adequately control the symptoms. Common sense, however, would dictate that a good diet and nutrition are suggested for the well being of any individual.

One substance, L-Tyrosine, which is an amino acid (protein), has been demonstrated to be effective in some cases. This natural substance is used by the body to synthesize (produce) norepinephrine (a neurotransmitter) which is known to be elevated by the use of amphetamines. Many new "natural" products have been recently introduced into the market as "cures" for Attention Deficit Disorder.

Psychological Treatment

Traditional child psychotherapy, such as play therapy or non-directive talking therapy, has not proven to be effective in the treatment of Attention Deficit Disorder nor has traditional family therapy. Providing individual psychotherapy for one or both of the parents also does not work. Research has demonstrated that modern psychological treatment methods, especially behavior modification, cognitive behavioral therapy and relaxation training can have a positive effect.

In some studies, one or several of these techniques used in combination have proven to be as effective as medication in reducing Attention Deficit Disorder symptoms. Counseling alone is used not to provide treatment but to provide education for the child and the family to help them better understand the disorder and how to cope with it. Counseling may also be used to help build self-esteem that has been damaged as a result of having Attention Deficit Disorder.

Modern psychological treatment methods can result in actual changes in mental functioning When there are changes in mental functioning there are corresponding changes in brain function. Changes in brain function then results in changes in brain metabolism. Thus, mental functioning and brain chemistry can be altered without the use of medication. Psychoeducational, behavioral and parent education interventions help children and teens with ADD/ADHD. Focus is a clinically proven, cost-effective program to provide these important treatment resources.

SOCIAL PROBLEMS

Some children with Attention Deficit Disorder experience significant problems socializing with peers and cooperating with authority figures. This is because when children have difficulty maintaining attention during an interaction with an adult, they may miss important parts of the conversation. This can result in the child not being able to follow directions and so called "memory problems" due to not listening in the first place. In this case, the child is not being disobedient or "strong willed" though they may be labeled as such. When giving directions to Attention Deficit Disorder children it is important to have them repeat the directions to make sure they have correctly received them. For younger children with Attention Deficit Disorder, the directions should consist of only one or two step instructions. For older children more complicated directions should be stated in writing.

Children with poor attention and concentration often miss important aspects of social interaction with their peers. When this happens, they have a difficulty time "fitting in." They need to focus in on how other the children are playing with each other and then attempt to behave similarly. Attention Deficit Disorder kids often enter a group play situation like the proverbial "bull in the china closet" and upset the play session. As they improve their ability to attend and concentrate, Attention

Deficit Disorder children can be coached on how to play appropriately with other children.

Attention Deficit Disorder children may have poor impulse control. This can result in several different problems during play time. First, they may have difficulty stopping a behavior once they have started. They may also carry the behavior to a level of intensity that is too much for the average child. This can even happen when the child is engaged in "horse play" with an adult. They often get "carried away" and don't know when to stop. This can result in negative feelings among those playing and make the others involved not want to play with the Attention Deficit Disorder child.

Sometimes an Attention Deficit Disorder kid will complain that when he gets into trouble at school that "all the other kids were doing the same thing and I was the only one that got in trouble." When you understand how an Attention Deficit Disorder child functions, it's possible to see how that may actually be close to the truth. Imagine that the teacher has left the room for a few moments. The class decides to take advantage of the situation and "mess around." When the teacher returns, the class sees her and they immediately stop what they are doing. On the other hand the Attention Deficit Disorder child may not immediately see the teacher enter the room and when he does is not able to immediately stop the inappropriate behavior. The teacher then reprimands him for not stopping.

Research shows that some children with Attention Deficit Disorder will eventually improve their ability to pay attention and concentrate whether they receive treatment or not. Their ability for self-control also improves. It is interesting that often children will require less medication as they get older. This may mean that the brain gradually is able to produce more of the needed neurotransmitters on its own without the help of medicine. However, some children seem to have increased difficulty with academic achievement around the fourth grade.

This may be due to the requirement for a higher level of organization and planning skills which they do not have. Because no one can predict how a child's Attention Deficit Disorder will progress, it is important that children who have difficulties with attention, concentration and impulse control, receive early intervention.

Psychological treatment such as that provided by Focus is most important because of it's lasting effects. Medication may also be useful in some cases. In the last few years, it has become more likely for adults to be diagnosed with Attention Deficit Disorder and start on medication. These may be individuals who were not properly identified when they were children. In other cases, adults may have learned to compensate for their disorder and become very successful individuals. In fact, some may "over compensate" and become extraordinarily successful.

Children with attention deficit disorder and/or learning disabilities can be a challenge for any classroom teacher. This page provides some practical suggestions that can be used in the regular classroom as well as the special education classroom. By looking through a given list of interventions, a teacher will be able to select one or more strategies that are suited to a specific child in a specific environment.

STRATEGIES FOR COGNITIVELY IMPULSIVE CHILDREN

Some children have difficulty staying with the task at hand. Their verbalizations seem irrelevant and their performance indicates that they are not thinking reflectively about what they are doing. Some possible ideas to try out in this situation include the following.

— Provide as much positive attention and recognition as possible.

— Clarify the social rules and external demands of the classroom.

— Establish a cue between teacher and child.

— Spend personal discussion times with these children emphasizing the similarities between the teacher and child.

— Get in a habit of pausing 10 to 16 seconds before answering.

— Probe irrelevant responses for possible connections to the question.

— Have children repeat questions before answering.

— Choose a student to be the "question keeper."

— Using a well known story, have the class orally recite it as a chain story.

When introducing a new topic in any academic area, have the children generate questions about it before providing them with much information. Distinguish between reality and fantasy by telling stories with a mix of fact and fiction and asking the children to critique them. Assign a written project that is to contain elements that are "true," "could happen but didn't," and "pretend, can't happen."

Helping Your Child With Arithmetic And Math

Children in trouble with arithmetic cannot seem to remember math problems even though they review them over and over again. They may seem to remember facts when reviewed on flash cards, but when presented with arithmetic problems, they must revert to finger counting or other aids to assist them. Following are some suggestions other parents have successfully used to help their children:

— Make sure your child can correctly write numerals. Even when children can count sequentially, they may have difficulties evidenced by reversing of numerals. Taking their hand in yours and tracing large numerals helps very much. Use a large, flat surface. Let your child get the "feet" of the shape. Try doing it with your child's eyes closed. Say the numeral as you trace it with him.

— If numeral reversals continue, help your child with the understanding Of "left" and "right" on his own body. Play games like "Loobie-Loo" that require moving one side of the body or the other. The awareness of left and right also affects letter reversals as well. [Click here for more suggestions on how to help children with reversals]

— Before and after games, with numbers, are helpful for math understanding. First, know how far your child can sequentially count. Then ask, "What number comes after ... ?" and "What number comes just before. . . ?" This skill is critical for understanding both addition and subtraction.

— Use numbers in a practical way around the house. "Susie, bring three forks to the table please;" or "Billy, will you give your dad five nails?" This gives children the opportunity to count in a realistic setting and to see, over and over again, that numerals in a problem at school represent real quantities. Use this activity in as many ways as you can.

— Board games, which involve tossing of dice or spinning that result in a number of moves across a board, are excellent ways to develop sequential math understanding. These games are particularly helpful if there are backward moves as "penalties" in the game. You can even let your child make his own game by using a large sheet of construction paper. Dominoes are a good math activity because, besides being a game, the matching of numbers (in the simple form of the game) is required. Children see the dots, can orally name them, and then can make the correct match.

— Keeping score on games played at home. There are any number of activities that children can do at home winch require tallying. Mom and Dad might play a game, and the child can record points by using the style of clustering four straight (upright) lines with the fifth running diagonally. Then, he can figure the totals by counting by fives.

— Give your child loads of opportunities to estimate space. This can be a family game if the conditions for involving other children are satisfactory. "How long do you suppose that table is?" Then it can be measured with a ruler or yardstick. The exact number of inches or feet is not critical. The question can be phrased so that the number of lengths is the critical factor. For example, "How many times would this ruler go across that table? You guess and I'll guess. Then we'll measure it. " You can practice estimating the distance across a room or up a wall, for example, in handprints, footsteps, paces, etc.

Measuring wall. Every home should have one wall that is used for keeping track of growth. Measure your child frequently and date each entry directly on the wall. Let him see how much he has grown as you measure him every month or every three months. The same thing can be done with plants. There are many bulb plants that grow quickly in a pot or jar. Put a ruler beside the container and let your child record the amount of growth each day. He can, keep a chart, with your help, to determine the daily growth.

Teaching children to tell time would be far simpler if training clocks had only an hour hand. If you happen to have a clock that Dad can take apart, remove the minute hand. Use a clock face with Arabic numerals. By using this dock, initially, and having it designated as "Johnny's clock," your child can see that it is "almost eleven," or "halfway between nine and ten; or, -a little after seven." When your child begins to understand words like "almost," "after," "in between," and how to use them, he will be ready to move to the two-hand clock.

Counting backwards is a game that children like because it ends with "Blast-off!" The skill of backwards counting is one that eventually develops the ability to understand subtracting by ones. It is also a visualization skill. Try starting from just "8" or "16" as practice. Count aloud with your child.

Counting and clustering real objects. Use beads or paper clips or buttons or poker chips-anything your child can grasp and that is not too law or too tiny. Let him arrange them into patterns or designs. Try clustering them into groups of two or three. Ask him for a specific number or trade items with turn.

Concentration. This game can be played in a number of ways. Generally, a specific number of playing cards are placed, face down, on the table. Your child turns a card over, one at a time, attempting to match two cards. The game calls for remembering where specific cards are placed, as he systematically searches for pairs. If he does not match a pair, cards are kept face down. Pairs are removed from the table. The game can be played with two people-or more.

"Fish" can also be played with playing cards. The object is to ask your opponent if he has a card you need to make a pair. Each player starts with four cards. Players take turns asking their opponent for a matching card. If the opponent does not have the "match," the asking player draws from the card stack. The game however, can be played as a multiplication game. Whatever pair is gotten, the child doubles or triples the face value of the cards.

Maintaining a daily calendar teaches, in an almost incidental way, adding by seven and multiplying by seven. Children can make their own calendars, with assistance, and then keep track of the passage of tune by crossing out each day after it has passed.

There are many ways of using division around the house if opportunities are used when they are available. In fact, creating them helps even more. Let your child assist you in separating things into even clusters. For example, after baking cookies, let your child assist you in solving the problem of how many should go into each place. As an incidental factor, mention, "That's right, twenty-one cookies and seven plates means each person gets three cookies-because 7 times 3 is 2 1. "

Mathematical, sequential reasoning enters into all kinds of daily uses. Determining halves, quarters, thirds, et cetera, when separating things is done daily in many households; for example, "Let's split this apple. You take half and I'll take the other half." Asking children to follow the directions involved in simple cooking activities gives them the opportunity to measure, mix, and follow a sequence to a natural conclusion.

Helping Your Child with Geography

There are certain kinds of information that youngsters need in order to function adequately as competent members of society. One of those areas is geography. Recent studies, even of college students, show that a surprising number of high school and college students have a remarkably poor understanding of geography-even a reasonable understanding of where major cities and countries are located in relation to one another.

Where to begin? Start early with a basic investment of at least four maps that can be hung in your child's room: a map of the world, the USA, your own state, and one of your local region. Check with your local library if you have any difficulty locating the maps you need.

If you are willing to go even one step further, invest in relief maps. These are particularly helpful when youngsters hear, for example, information such as "The Mississippi River basin drains two/thirds of the United States." By noting the location of the Rocky Mountains on the western third of the continent and the contours of the eastern slopes of the United States, the child can readily see why this fact is true. On such maps it is also obvious why populations are concentrated more in some areas of the country than others and why parts of the country lend themselves more to farming and ranching than industrialization, as well as the importance of ports and river ways, and so on.

Once the maps are mounted, identify the four major points of the compass-north, east, south, and west Make large, clear labels for these and put them on the maps. (A note-the top of a map is always north.)

Basic facts about the local area should be taught. Your child should know adjoining or nearby cities and that they are "to the east of' or "north of" his city of residence. With the massive highway connections in this country, youngsters can "fix" some locations by their proximity to major thoroughfares.

Virtually a day doesn't pan that some event of consequence isn't reported from some place on the globe. These countries and cities can be marked with little flags.

With many families separated because of jobs and other factors, a map can be used to show San Francisco where Aunt Susan fives, or Tallahassee where Uncle Jim fives, or Washington, DC, where Grandma lives, or Boston where Cousin Ed works. An effort should be made to give the child a sense of key spots that can serve as "anchors"-for example, "Connecticut? That's the state next to Massachusetts where Cousin Ed works."

Next to the bathroom, the kitchen is the most commonly used room in the house. Don't overlook placing a large map on one wall. As news of consequence about problems in specific parts of the world turmoil, earthquakes, floods, monsoons, famine, oil-arise, their locations can be flagged for everyone's benefit.

Every home should have an Atlas of fairly recent date-one that, like a dictionary, isn't stored on a shelf collecting dust but is used for frequent reference. An atlas makes a noteworthy birthday or holiday gift. In addition to the locations of countries and cities, a tremendous amount of other important information is given-products, resources, populations, languages spoken, and so on.

If your child really gets turned on to maps, introduce the concept of "scale. " Every map has a scale that shows how many inches equal a mile. The smaller the map, the more miles will be squeezed into an inch. Get out a ruler when figuring distance and ask your child to determine the approximate number of miles from one place to another. This activity can take on added meaning if the family is planning a trip, and the number of miles involved is important. (If you really want to get your child involved and interested, tell him how many miles to the gallon your car uses in highway travel. See if he can figure out the number of gallons of gas that will be needed for the trip. He can then multiply that figure by the price of a gallon of gas to estimate the cost of the trip.)

Make good use of your local area map. What are spots of interest to your child-places he knows or likes to go, such as the football field, the city hall, the city park, the zoo, the sports arena, the civic center? Mark these with "flags" to relate them to "Flag #1" (your home). If time permits (and often it doesn't in busy households in the morning families prepare for school and work) the morning news shows on the major television stations show a considerable number of news, political and weather drawings and maps. Just from repetition alone, a certain amount of geographical information will "stick" aside from the other valuable and timely information.

There are puzzles of maps of the United States and the world that add yet another dimension to becoming familiar with our country and world. These range from very simple to complex and should be selected based on the child's current ran & of knowledge-too easy would be insulting; too difficult would be frustrating. If a puzzle of a map at the proper level cannot be found, it is possible to make one at home by mounting a map on a stiff piece of cardboard and cutting it into puzzle pieces.

In some commercial maps of the United States and the world, the actual shape of the state or country forms the pieces. This is an excellent way for the child to learn the size and shape of his own state and country and how they relate in size to other states and other countries. Just memorizing states and their capitals doesn't teach geography. Such information has to be linked into other facts-facts that have special meaning for the child.

One thing all youngsters should do, either as a government class project or with the family, is to visit their own state capital and arrange to visit legislative sessions-and to contact in advance their local state representatives. This is one sure way that they will team and remember the name of their state capital and its location relative to other cities. But a trip to the state capital should involve a bit of advance study. Along the way, whether it is by car, train, or bus—or even airplane—certain key spots (rivers, historical locations, and cities) will have to be passed. Knowing these in advance will "lock in" additional geographical information.

On a family outing, keep a record of mileage and time so that your child understands the distance involved. If you are taking a trip that might involve more than an hour, record the starting and ending mileage. A great source of geographical information that is bright, colorful, and well designed is any local Chamber of Commerce. A phone call or letter requesting printed materials will unfailingly bring a response. Another source is your local travel agency whose representatives will be glad to share their information brochures.

Almost every major airline has a fine magazine tucked into the pocket in front of the seat. These publications are a treasure of geographical information, including maps that show air routes. If a family member or a friend is scheduled to make an air trip, ask that person to bring back one of these magazines. Another valuable resource is an automobile club. If you or a family member or a friend is a member, check out their ample supply of maps, both local and otherwise. Such clubs will also provide "strip maps" with routes plotted which are models of clarity and will interest and fascinate the child.

While knowledge of the locations of various places around the globe is almost second nature to most adults it can occasionally be perplexing and confusing to children. Stimulating talk about the news and why, for example, an early frost in Brazil will raise the price of coffee, or how difficulties in the Middle East can affect the price of oil, or why a strike in the automobile industry in Michigan may affect the cost of can will, little by little, help youngsters get a better grasp and sense of geography.

Helping Your Child with Handwriting

Children who paint or write in cursive, but who are unable to write legibly and consistently, in spite of repeated admonitions, require special approaches to the solution of their special difficulties. These are youngsters who are unable to properly form their letters, who have difficulty keeping their letters on the line, who may not seem to understand the relative sizes of letters, who either crowd letters within words together, or who space so poorly that it is almost impossible to determine where one word ends and another begins. The net result is that what they have written is often difficult or near impossible to decode, even when it is spelled correctly.

Finger painting is a messy activity unless you have a law area that won't be too difficult to clean. Oil cloth on an old table or on concrete or vinyl floor works quite well. Use a plastic apron on yourself and your child. Have him roll the paint around in huge circles so that not only his hands, but his elbows and shoulders are involved. Just playing with shapes on the slippery surface helps tremendously. Making shape designs is fun and reinforces the development of shape constancy.

When children just can't seem to stay "on the line" as they print or write, try using a red felt tip pen to rule across the lines that will be the bottoms of letters. You may also want to use a green felt tip pen just to remind your child when to begin his strokes, since printed letters start basically at the top and go down. Clay can be purchased from crafts stores in twenty-five pound sacks, often for under $5.00. Letting children mold the clay into forms gives them another kind of experience with shapes, but in a three- form that is helpful for form recognition. They can also form "snakes" and make letters, even their own names.

Quite often children hold pencils and crayons in an awkward manner and grasp. To develop the strength in the hands and fingers for proper grasp, let your child do activities that require holding or hanging. Make good use of your school play yard. Let him hang by his hands from the jungle gym to develop strength in the shoulder girdle as well as his hands. Squeezing objects, such as little rubber balls, or playing with wooden clothespins help to develop finger coordination and strength.

One of the prerequisites for handwriting is the ability of the eyes to work in close cooperation with the hands. This means that the eyes themselves must be able to move smoothly and must be able to follow moving targets. General motor coordination (balancing, hopping, running, skipping, et cetera) is necessary for laying the groundwork for smooth, fine muscle control. Play, for example, flashlight tag with your child. This requires two flashlights and a dark room. You be "It" and see if your child can, with his flashlight, "tag" your light.

Play tracing games. Have your child sit next to you with his eyes closed. Take his writing hand, index and middle fingers pointing and the other fingers flexed, and a-ace a shape or letter on a large surface. See if he can guess what shape or letter you traced.

If you're prepared to be squirted, and it's a warm day, and your back yard has a sunny wall, try this one. Get a squirt gun and let your child "write" letters with water on the wall. The sun will dry the letters reasonably fast. This allows your child to use space and estimate, on a large surface, just how he will execute the proper formation of the letter.

Observe the way your child sits when he writes. As a check, try this yourself. Sit at a table so that your elbows comfortably rest on the surface. Then fold your hands in front of you, flat on the desk so that your body and folded hands form a triangle. If you are right-handed, the paper would go directly under that folded arm. If you are left-handed, the paper would go directly under that folded arm. Notice that when you old the pencil, after this experiment, that the writing hand touches the surface of the paper directly along the line of the little finger and wrist. If you are right-handed, your back and head will be slightly curved to the left. If your child is doing anything other than this, it means that he is not ready for the activity, or it is too demanding for him. It may also suggest that he has visual difficulties in the way he uses his eyes.

If a child continues to reverse letters, even as his handwriting improves, give him opportunities to identify left and right on his own body. Play games requiring use of just the left hand or the right hand or the left foot or the right foot. Play "blind man's bluff, in which you must direct him across a room by giving him turns to make. Have him direct you when it's your turn. If you notice that your child continually holds his pencil right at the tip, it ran suggest that too much pressure is required for holding it properly. Try using a rubber band, twisted several times, and place it just above the shaved area. This will provide a tactile reminder on where to hold it.

"Rhythmic writing" is a term applied to law handwriting at a chalkboard. On the chalkboard you've made for home use, have your child stand so that he is facing the center of the board. Then, if he is right-handed, have him start a series of "e" letters, all connected, and all moving from left-to-right. As he moves from left-to-right with his writing hand, he should keep his feet firmly planted in one spot, and move his arms as far as he can. Then he can practice with "y" letters, and then combine "e" and "y" across the board.

If you have a large sink area with a formica top, carefully "Soap" it. Don't make it too wet or you'll have a mess in the kitchen. Let your child stand by it and practice writing his letters, one at a time. Again, it is good for getting the "feel" of the letters. You can also take his hand, as in finger-painting, and move it through the slick surface, to form specific letters that are difficult for him.

Encourage your child to use what he learns. Go on a sign-making spree. Let him write (and decorate) signs that say, for example, "This is Jimmy's room. Enter at your own risk," et cetera. He can help you prepare a shopping list or birthday list. You'll undoubtedly have dozens of ways your child can use his developing skill in a practical way. Play games with plastic letters that can be purchased at most local variety and school supply houses. These come in two forms both manuscript-upper (capitals) case and lower (small letters) case.

In order to print a letter a child must be able to visualize the shape of the letter. Let your child take one of the plastic letters and feel it with his eyes dosed. Can he recognize and name it? Can he draw it even if he is unable to name it? Let him describe it as he is feeling the surface and the sides. On confusing letters such as "h" and "n," which many children have difficulty with, let him feet them, one at a time, and help him feet the difference between the two.

When a child develops proper formation of letters, particularly in cursive, but does not maintain a constant slant, try this. Even though it takes a little time, it is worth it. With a ruler, pencil-in diagonal lines, very lightly, across the paper. These diagonal lines should be carefully done so that they provide "guidelines" for your child. As he writes, he has a visual set of "clues" to use to make sure his letters all slant the same way.

Keep in touch with your child's teacher as your youngster works at home with you to develop his skills in handwriting. Try not to make your child feel that he isn't "trying hard enough" or that you "just can't read it, it's so bad. Words of encouragement go a long way with children, just as they do with adults, and they are truly a significant part of any home activity that is designed to help a child.

Kids and Teens With Homework and Study Habits

Certain key practices will make life easier for everyone in the family when it comes to study time and study organization. However, some of them may require an adjustment for other members of the family. For lots of helpful internet tools for research and mastering subjects visit our Homework Help Center.

Turn off the TV set. Make a house rule, depending on the location of the set, that when it is study time, it is "no TV" time. A television set that is on will draw youngsters like bees to honey.

What about the radio? Should it be on or off? Contrary to what many specialists say, some youngsters do seem to function all right with the radio turned on to a favorite music station. (Depending on the layout of your house or apartment, maybe an investment in earphones would be worthy of consideration.)

Certain rules should be set about the family phone during study hours. The more people in the household, the more restrictions on long and unnecessary phone calls are needed. A timer, placed next to the phone, can help to control the length of calls so that the telephone will be available if it becomes necessary to call a schoolmate to confirm an assignment or discuss particularly difficult homework.

Designate specific areas for homework and studying. Possibilities include the child's room or the kitchen or dining room table. Eliminate as much distraction as possible. Since many young people will study in their own rooms, function becomes more important than beauty. Most desks for young people really don't have sufficient space to spread out materials. A table that allows for all necessary supplies such as pencils, pens, paper, books, and other essentials works extremely well.

Consider placing a bulletin board in your child's room. Your local hardware store sells wallboard that might not look too pretty and isn't framed, but a 4 x 3'section is inexpensive and perfect on which to post pertinent school items. You might want to paint or cover it with burlap to improve its appearance or let your child take on this project.

Encourage the use of a small book or pad for writing down assignments so that there is no confusion about when certain assignments must be turned in to the teacher.

Keeping general supplies on hand is important. Check with your child about his needs. In fact, make it his responsibility to be well supplied with paper, pencils, note pads, notebook paper, et cetera.

Regularity is a key factor in academic success. Try to organize the household so that supper is served at a standard time, and once it and family discussions are over, it's time to crack the books. If the student doesn't have other commitments and gets home reasonably early from school, some homework can be done before supper.

Consider you child's developmental level when setting the amount of time for homework. While high school students can focus for over an hour, first-graders are unlikely to last more than 15 minutes on a single task. Allow your child to take breaks, perhaps as a reward for finishing a section of the work.

Organize study and homework projects. Get a large calendar, one that allows space for jotting down things in the daily boxes. Rip it apart so that you (and the child) can sequentially mount the school months for the current semester. For example, you can tear off September, October, November, December, and January and mount them from left to right across one wall. Have the child use a bold color writing instrument (felt tip pen) to mark exam dates in one color, reports that are coming due in a different color, et cetera. This will serve as a reminder so that things aren't set aside until the last dangerous moment.

Teach your child that studying is more than just doing homework assignments. One of the most misunderstood aspects of schoolwork is the difference between studying and doing homework assignments. Encourage your child to do things like:

— take notes as he's reading a chapter

— learn to skim material

— learn to study tables and charts

— learn to summarize what he has read in his own words

— learn to make his own flashcards for quick review of dates, formulas, spelling words, et cetera

Note-taking is a critical skill and should be developed. Many students don't know how to take notes in those classes that require them. Some feel they have to write down every word the teacher says. Others have wisely realized the value of an outline form of note-taking. Well prepared teachers present their material in a format that lends itself to outline form note taking..

In some cases, they should be, particularly if a lot of material was covered, and the youngster had to write quickly but lacks speed and organization. A home dictionary is essential, but if it is kept on a shelf to gather dust, it won't do anyone any good. Keep it in an accessible place and let your child see you refer to it from time to time. If the family dictionary is kept in the living room and the child studies in his room, get him an inexpensive dictionary for his exclusive use.

Good dictionary, encyclopedia and organizational skills depend on the ability to alphabetize. See if your child's teacher practices alphabetizing in class. Try alphabetizing spelling words, family members' names or a few favorite toys at home as a way of practicing. Help your child to feel confident for tests. Taking tests can be a traumatic experience for some students. Explain to your child that burning the midnight oil (cramming) the night before a test is not productive. Better to get a good night's sleep.

Students also need reminding that when taking a test, they should thoroughly and carefully read the directions before they haphazardly start to mark their test papers. They should be advised to skip over questions for which they don't know the answers. They can always return to those if there's time. Good advice for any student before taking a test: take a deep breath, relax, and dive in. Always bring an extra pencil just in case.

During a homework session, watch for signs of frustration. No learning can take place and little can be accomplished if the child is angry or upset over an assignment that is too long or too difficult. At such times the parent may have to step in and simply halt the homework for that night, offering to write a note to the teacher explaining the situation and perhaps requesting a conference to discuss the quality and length of homework assignments.

Should parents help with homework? Yes-if it is clearly productive to do so, such as calling out spelling words or checking a math problem that won't prove. No-if it is something the child can clearly handle himself and learn from the process. And help and support should always be calmly and cheerfully given. Grudging help is worse than no help at all!

Read directions, or check over math problems after your child has completed the work. Remember to make positive comments - you don't want your child to associate homework with fights at home. Model research skills by involving your child in planning a family trip. Help your child locate your destination on a map or atlas. Use traditional encyclopedia or a CD-ROM to find information about the place you will visit; try the Internet or books in the library.

How best to handle report cards? To save shocks and upsets, gently discuss from time to time "how things are going at school- with your child. Something casual, such as "How did the math test go?" "How did you do on the history report?" "How's your science project coming along? Need any help?" are questions that aren't "third degree" but indicate interest. Find out if it is a policy at your child's school to send out "warning notices" when work isn't going well.

Generally, such notices require the parent's signature to verify that the parent has, indeed, been alerted. This is the time to contact the teacher of the course, along with your child, to learn what the difficulty may be. If such notices aren't sent, then grades on projects and reports and from tests may be the sole source of information short of what your child wishes to share. Be tuned in to statements such as "He's an awful teacher," "She goes too fast," etc. This may be the child's way of indicating frustration in understanding content or lack of study time with the subject. However, be cautious in contacting teachers without your child's approval or interest. It may disrupt good feelings between you and make you seem to be interfering and spying.

Establishing a Parent-Teacher Relationship

The first contact with your child's teacher, in many ways, is the most important, This is the time you are building rapport and developing a relationship of trust. Therefore, an appropriate time and setting is important for the first brief encounter. A phone call, a note, or, best of all, an initial face-to-face meeting is best. A good time to contact your child's teacher is during the first week of school. This gives you an opportunity to meet one another when neither has any complaints. Otherwise, the first teacher contact can be unpleasant.

The teacher is usually calling to describe some unacceptable behavior or report a child's tack of progress and her concern that a learning problem may exist. This kind of contact usually puts a parent on the defensive, and communication can be hampered. Neither party wins, and the biggest loser is your child. However, during the first week of school, the teacher probably knows very little about your child. Thus, you are in a position to provide some helpful information. This is the time to mention then identify these. And, last but not least, assure the teacher that she has your full support and cooperation. Provide the teacher with your phone number and tell her to feel free to call when help is needed from home. Let the teacher know from the start that you want to work with her, not against her, so your child will learn. Do not feel you are intruding or asking for special treatment. You are simply indicating that you are truly concerned that your child receives a good education.

After your child has spent six weeks in school, again call or drop a note to check on your child's progress. If a conference needs to be set up, do it immediately. Even if your child is doing well, you may still want a conference.

GUIDELINES

Let's consider specific guidelines to help you communicate effectively with your child's teacher. Practice these guidelines, and your child will reap the benefits.

Guideline 1: Identify the purpose for the conference. Is it to become acquainted? Is it to alleviate your concerns about your child's poor attitude towards reading and/or school! Or is it to receive a report card and test scores? Each of these situations is vastly different and requires different preparation.

Guideline 2: Communicate the purpose for the conference. If you are requesting the conference, immediately tell the teacher the purpose. This helps to alleviate any imagined fears the teacher may have about your request to hold a conference.

Guideline 3: Arrange the conference at the teacher's convenience. The teacher then has sufficient time to plan and to have the necessary information at the conference. An unplanned conference can turn out to be a waste of time for both teacher and parent and cause feelings of frustration.

Guideline 4: Plan for the conference. Write out the areas and questions you want the conference to cover. Combine, delete, and clarify these questions, and, finally, prioritize them. By using this process, your most important questions will be answered in a clear, succinct manner. Moreover, the teacher's responses will likely be clearer and more to the point.

Guideline 5: Restate the purpose of the conference at the onset. Try to stay a on one topic since your time together is limited.

Guideline 6: Display a positive attitude during the conference. Be aware that not only what you say reflects your attitude, but also your tone of voice, facial expression, and body movements. A loud voice may imply dominance. Rigid posture may suggest anger or disapproval. Always listen attentively and show your enthusiasm.

Guideline 7: Remain open and supportive throughout the conference. Don't become antagonistic or defensive; otherwise the conference outcome can be disastrous. Strive for cooperation between you and your child's teacher. Even when teachers present a negative side of your child's behavior or inform you of other problems, try to remain objective. This can be difficult when it is your child, but he will experience as many or more difficulties if you and die teacher don't try to find a way to work together to solve these problems.

Guideline 8: Make sure suggestions are provided to increase your child's growth. If your child is doing well, find out what you can do to ensure continued success and progress. If he has difficulties, make sure the teacher goes beyond merely pointing out a problem. The teacher needs to provide ideas for eliminating or reducing the difficulty. Many parents have been discouraged or aggravated because teachers point out problems but don't provide solutions. Don't let this situation occur! If immediate suggestions can't be provided, then a follow-up conference is needed.

Guideline 9: Ask for examples of daily work to better understand your child's strengths and weaknesses By reviewing your child's work, you will learn if progress has been made since the last conference. Have any weaknesses become more severe? If improvement hasn't been made, are other methods or materials being used? As a parent, what should you be doing at home with your child?

Guideline 10: Clarify and summarize each important point as it is discussed. Thus, both teacher and parent are better able to develop a mutual understanding and agreement. Let's took at a conference in which a parent does a good job of clarifying and summarizing a major point.

Guideline 11: Owe agreement is reached, discuss the next topic. During the conference, you may want the teacher to understand certain things about your child. or you may have a special request. Once your point is understood and the teacher has agreed, it is wise not to continue the same discussion. It may present new questions which may reverse the previously made agreement. Once a decision is made, it is best to start discussing the next point. You will find the conference to be much more productive.

Guideline 12: Make sure you understand the information the teacher is supplying. Often teachers use educational jargon, not realizing parents don't understand. Don't be afraid to ask for an explanation or definition. Make sure when the conference ends you have understood all the information reported. if you're confused or uncertain, your child won't benefit and learning may be hindered.

Guideline 13: Keep conferences short. Conferences that run more than 40 minutes can be tiresome for both parent and teacher. If you can't accomplish all that has been planned, ask for another conference. By scheduling a future conference, you will have an opportunity to follow up on previous agreements and revise them if necessary

READING PROBLEM

Reading is not automatic but must be learned. The reader must develop a conscious awareness that the letters on the page represent the sounds of the spoken word. To read the word "cat," the reader must parse, or segment, the word into its underlying phonological elements. Once the word is in its phonological form, it can be identified and understood. In dyslexia, an inefficient phonological module produces representations that are less clear and, hence, more difficult to bring to awareness.

The main reasons for reading problems are:

— Ineffective reading instruction

— Auditory perception difficulties

— Visual perception difficulties

— Language processing difficulties

Helping Your Child With Spelling

Children who fall behind classmates in spelling, who forget words easily, or who mix up letters when writing, are children who need special, loving, considerate attention regularly at home to help them overcome their unique learning problems. Forcing an activity on them or making it over-demanding only serves to intensify the child's negative feelings about it. Here are some suggestions other parents have successfully used to help their children in these areas.

Develop auditory and vocal skills. Good spellers are usually good readers and good speakers and vice versa. Using the school or local library helps your child develop some of these skills. Restrict the amount of TV he watches. Your child cannot talk back to a television set. Children need to use the language they will be writing. Give your child the opportunity to talk with you.

Experience stories. Let your child write about the things he likes. He can illustrate the stones himself or cut pictures from magazines to illustrate them. Let him write the words without assistance unless he asks for help. Misspelled words can then be used in little games you play with your child. A one-line "story" may be all that he can handle. If so, fine.

Write letters. Corresponding with a friend or relative-or a simple statement at the bottom of a letter you write to someone your child knows-offers him opportunities to spell.

Trace words. This activity helps many children. Have your child sit next to you (or, if young, sit on your lap). Sit so that you can guide his writing hand. Make sure that only his index and middle fingers are extended, and that his eyes are closed. If your child is using manuscript, use that form. Take his hand and print (or write) the word that is confusing to him.

Finger paints are messy, but ever so helpful. Use oilcloth and a large table. Have your child roll up his sleeves and wear an old apron. Let him use both hands to write letters and words, It is a marvelous activity. Just gating the feel of large movements may be sufficient without introducing formal spelling to the activity.

All kids love codes, so why not encourage your child to decode messages diet you leave for him? Let him make up his own codes for you. You make up one but make sure he has a way to decode it.

If your child is working on a class spelling list and can only remember half of the words, speak to his teacher. Teachers are more than delighted to hear how their students respond to homework. Perhaps the list can be reduced so that your child has fewer words and can learn these more efficiently and comfortably.

Don't tackle an entire spelling bar in one sitting. Take one-third for example, each evening, to work on with your child. Break the practice into small units. Try fifteen minutes of review when he gets home; fifteen minutes before supper; fifteen minutes after supper. Shorter periods given frequently are more effective than one massive review-which is also exhausting and frustrating.

Sometimes words on a spelling list can be "clustered" into similarities. For example, you might try attempting all of the five-letter words one day, all the words beginning with consonants the next day, all the words beginning with blends the next day. This kind of grouping will help your child to perceive similarities and differences in the words, and, hence, develop his recall.

An old trick that really works is to have your child practice. Write each of his words, and then draw with a black crayon around each word. Then he can lightly color the shape of the word. This is known as "studying the shape" (or configuration). Just make sure your child uses straight, not curved, lines when he outlines the word.

Before your child starts to silently study his list for that day, let him pronounce each word. Children must know how to properly pronounce a word before they attempt to spell it, If their pronunciation is not correct, they will indeed spell it as they would pronounce it in their own way. (Also make sure they know what the word means and can use it or understand it when they hear it.)

To start studying, a child should look at the word, pronounce it, spell it orally as he looks at it, cover it with his hand, and then attempt to spell it or, as he traces it on your kitchen table, letter by letter.

After your child has studied, let's say five words, in the manner described, spell the words to him, in random order, and have him name the word you spelled.

Invest in a set of plastic magnetic letters that are available at many discount, toy, and variety stores. Let your child spell the word by successively placing the magnetic letters on the magnetic board. You can show him the word, then remove it Have him name each letter as he locates it and places it on the board. This is good for developing the correct order for letters within the words.

Word lists. These can be made using paper available in the house. Print or write the words being studied. Post one copy of the list on die refrigerator, another on the door to your child's room, and another in the bathroom. Maybe another can go over the TV set. Use a different color crayon for each word-or use a different color for parts of each word regularly confusing your child. For example, if he continues to write "come" as "cum," use black for the "c" and I'm" but red for the "o" and "e".

Put movement into learning words. Have your child clap for each letter or take a step for each letter as he spells the word orally. This will help " lock in" the correct sequence of letters, as well as develop full recall for the word.

Let your child play teacher. Let him teach you the words he is learning to spell. Spell them orally to him. Let him correct you. Then have him dictate to you and you write them. Have him score your paper. Make a game of it. He'll know you really know how to spell them, so veil him it's a game.

Commercial dice with letters rather than numbers. Take turns with your child in tossing them and building words. List the words as they are made. The list can be saved and added to each time you play. That way be can develop a "reference fist" to use over and over spin to reinforce his recall.

Listening skills do help spelling. "What letter does 'chart, end with?" What letter does 'piano' begin with?" Play these games just for a few moments before supper, or after breakfast to develop your child's ability to hear sounds in words.

Rhyming words is another game that can build spelling skills. "Can you think of a word that rhymes with fill?" As your child says hill, Bill, till, and so on, write them down. He'll soon notice, himself, that they have identical endings.

Remembering. This is a game to develop visual memory. Write one word on a piece of paper. Leave space underneath it. Tell your child to look at it as long as he wants, that is, until he can remember the letters, then have him fold the paper so that he cannot see the word. He is then to try to write the word from memory. Let him check it himself, and if he has misspelled, try again.

Practice in spelling can come in a variety of ways. For example, you might ask him to help you make a grocery list by looking at the advertisement for a local supermarket You could check the items you want to purchase, and you could ask him to make a list to help you out.

Find the wrong word. Write a short sentence for your child. Tell him that there is one word spelled incorrectly. Ask him to see if he can find it. To begin, make it a rather obviously misspelled word. Leave a letter out, or add an extra letter to a word. Ask him to first read the sentence, then to circle the misspelled word. Then make sure you erase it and write it correctly.

Helping your child at home with spelling requires patience and a non-school-like setting. Don't try to be a teacher. Be a parent who teaches. Your child wants to please you-he wants your honest, sincere praise. Keep the activities short-and fun, and do them regularly, with variety.

5

Emotional Development of Children

According to Erikson, the socialization process consists of eight phases - the "eight stages of man." His eight stages of man were formulated, not through experimental work, but through wide - ranging experience in psychotherapy, including extensive experience with children and adolescents from low - as well as upper - and middle - social classes. Each stage is regarded by Erikson as a "psychosocial crisis," which arises and demands resolution before the next stage can be satisfactorily negotiated. These stages are conceived in an almost architectural sense: satisfactory learning and resolution of each crisis is necessary if the child is to manage the next and subsequent ones satisfactorily, just as the foundation of a house is essential to the first floor, which in turn must be structurally sound to support and the second story, and so on.

ERIKSON'S EIGHT STAGES OF DEVELOPMENT

1. *Learning Basic Trust Versus Basic Mistrust* (Hope):Chronologically, this is the period of infancy through the first one or two years of life. The child, well - handled, nurtured, and loved, develops trust and security and a basic optimism. Badly handled, he becomes insecure and mistrustful.

2. *Learning Autonomy Versus Shame* (Will): The second psychosocial crisis, Erikson believes, occurs during early childhood, probably between about 18 months or 2 years and 3½ to 4 years of age. The "well - parented" child emerges from this stage sure of himself, elated with his new found control, and proud rather than ashamed. Autonomy is not, however, entirely synonymous with assured self - possession, initiative, and independence but, at least for children in the early part of this psychosocial crisis, includes stormy self - will, tantrums, stubbornness, and negativism. For example, one sees may 2 year olds resolutely folding their arms to prevent their mothers from holding their hands as they cross the street. Also, the sound of "NO" rings through the house or the grocery store.

3. *Learning Initiqtive Versus Guilt* (Purpose): Erikson believes that this third psychosocial crisis occurs during what he calls the "play age," or the later preschool years (from about 3½ to, in the United States culture, entry into formal school). During it, the healthily developing child learns:

   (1) to imagine, to broaden his skills through active play of all sorts, including fantasy

   (2) to cooperate with others

   (3) to lead as well as to follow.

   Immobilized by guilt, he is:

   (1) fearful

   (2) hangs on the fringes of groups

   (3) continues to depend unduly on adults and

   (4) is restricted both in the development of play skills and in imagination.

4. *Industry Versus Inferiority* (Competence): Erikson believes that the fourth psychosocial crisis is handled, for better or worse, during what he calls the "school age," presumably up to and possibly including some of junior high school.

Here the child learns to master the more formal skills of life:

(1) relating with peers according to rules

(2) progressing from free play to play that may be elaborately structured by rules and may demand formal teamwork, such as baseball and

(3) mastering social studies, reading, arithmetic.

Homework is a necessity, and the need for self-discipline increases yearly. The child who, because of his successive and successful resolutions of earlier psychosocial crisis, is trusting, autonomous, and full of initiative will learn easily enough to be industrious. However, the mistrusting child will doubt the future. The shame - and guilt-filled child will experience defeat and inferiority.

5. *Learning Identity Versus Identity Diffusion* (Fidelity): During the fifth psychosocial crisis (adolescence, from about 13 or 14 to about 20) the child, now an adolescent, learns how to answer satisfactorily and happily the question of "Who am I?" But even the best - adjusted of adolescents experiences some role identity diffusion: most boys and probably most girls experiment with minor delinquency; rebellion flourishes; self - doubts flood the youngster, and so on.

   Erikson believes that during successful early adolescence, mature time perspective is developed; the young person acquires self-certainty as opposed to self-consciousness and self-doubt. He comes to experiment with different - usually constructive - roles rather than adopting a "negative identity" (such as delinquency). He actually anticipates achievement, and achieves, rather than being "paralyzed" by feelings of inferiority or by an inadequate time perspective.

   In later adolescence, clear sexual identity - manhood or womanhood - is established. The adolescent seeks leadership (someone to inspire him), and gradually develops a set of ideals (socially congruent and desirable, in the case of the successful adolescent). Erikson believes that, in our culture, adolescence affords a "psychosocial moratorium," particularly for middle - and upper-class American children. They do not yet have to "play for keeps," but can experiment, trying various roles, and thus hopefully find the one most suitable for them.

6. *Learning Intimacy Versus Isolation* (Love): The successful young adult, for the first time, can experience true intimacy - the sort of intimacy that makes possible good marriage or a genuine and enduring friendship.

7. *Learning Generativity Versus Self-Absorption* (Care): In adulthood, the psychosocial crisis demands generativity, both in the sense of marriage and parenthood, and in the sense of working productively and creatively.

8. *Integrity Versus Despair* (Wisdom): If the other seven psychosocial crisis have been successfully resolved, the mature adult develops the peak of adjustment; integrity. He trusts, he is independent and dares the new. He works hard, has found a well - defined role in life, and has developed a self-concept with which he is happy. He can be intimate without strain, guilt, regret, or lack of realism; and he is proud of what he creates - his children, his work, or his hobbies. If one or more of the earlier psychosocial crises have not been resolved, he may view himself and his life with disgust and despair.

These eight stages of man, or the psychosocial crises, are plausible and insightful descriptions of how personality develops but at present they are descriptions only. Helping the child through the various stages and the positive learning that should accompany them is a complex and difficult task, as any worried parent or teacher knows. Search for the best ways of accomplishing this task accounts for much of the research in the field of child development.

Socialization, then is a learning - teaching process that, when successful, results in the human organism's moving from its infant state of helpless but total egocentricity to its ideal adult state of sensible conformity coupled with independent creativity.

TEMPERAMENT AND CHILD'S PERSONALITY

Personality is determined by the interaction of temperament traits with the environment. Each person (including your child) comes with a factory installed wiring. How your child is wired can determine whether they will be easy or difficult to raise. How well their temperament fits with the environment and how well they are received by the people in the environment will determine how a child sees himself and others.

What is temperament?

Temperament is a set of in-born traits that organize the child's approach to the world. They are instrumental in the development of the child's distinct personality. These traits also determine how the child goes about learning about the world around him. These traits appear to be relatively stable from birth. They are enduring characteristics that are actually never "good" or "bad." How they are received determines whether they are perceived by the child as being a bad or good thing.

When parents understand the temperament of their children, they can avoid blaming themselves for issues that are normal for their child's temperament. Some children are noisier than other. Some are more cuddly than others. Some have more regular sleep patterns that others. When parents understand how their child responds to certain situations, they an learn to anticipate issues that might present difficulties for their child. They can prepare the child for the situation or in other cases they may avoid a potentially difficult situation all together.

Parents can tailor their parenting strategies to the particular temperamental characteristics of the child. They can also avoid thinking that a behavior that reflects a temperament trait represents a pathological condition that requires treatment. Parents feel more effective as they more fully understand and appreciate their child's unique personality.

When the demands and expectations of people and the environment are compatible with the child's temperament there is said to be a "goodness-of-fit." When incompatibility exists, you have what is known as a "personality conflict." Early on parents can work with the child's temperamental traits rather than in opposition to them. Later as the child matures the parents can help the child to adapt to their world by accommodating to their temperamental traits.

The 9 temperamental traits are;

1. *Activity Level*: This is the child's "idle speed or how active the child is generally. Does the infant always wiggle, more squirm? Is the infant difficult to diaper because of this? Is the infant content to sit and quietly watch? Does the child have difficulty sitting still? Is the child always on the go? Or, does the child prefer sedentary quiet activities? Highly active children may channel such extra energy into success in sports; may perform well in high-energy careers and may be able to keep up with many different responsibilities.
2. *Distractibility*: The degree of concentration and paying attention displayed when a child is not particularly interested in an activity. This trait refers to the ease with which external stimuli interfere with ongoing behavior. Is the infant easily distracted by sounds or sights while drinking a bottle? Is the infant easily soothed when upset by being offered alternate activity? Does the child become sidetracked easily when attempting to follow routine or working on some activity? High distractibility is seen as positive when it is easy to divert a child from an undesirable behavior but seen as negative when it prevents the child from finishing school work.
3. *Intensity*: The energy level of a response whether positive or negative. Does the infant react strongly and loudly to

everything, even relatively minor events? Does the child show pleasure or upset strongly and dramatically? Or does the child just get quiet when upset? Intense children are more likely to have their needs met and may have depth and delight of emotion rarely experienced by others. These children may be gifted in dramatic arts. Intense children tend to be exhausting to live with.

4. *Regularity*: The trait refers to the predictability of biological functions like appetite and sleep. Does the child get hungry or tired at predictable times? Or, is the child unpredictable in terms of hunger and tiredness? As grown-ups irregular individuals may do better than others with traveling as well as be likely to adapt to careers with unusual working hours.

5. *Sensory Threshold*: Related to how sensitive this child is to physical stimuli. It is the amount of stimulation (sounds, tastes, touch, temperature changes) needed to produce a response in the child. Does the child react positively or negatively to particular sounds? Does the child startle easily to sounds? Is the child a picky eater or will he eat almost anything? Does the child respond positively or negatively to the feel of clothing? Highly sensitive individuals are more likely to be artistic and creative.

6. *Approach/Withdrawal*: Refers to the child's characteristic response to a new situation or strangers. Does the child eagerly approach new situations or people? Or does the child seem hesitant and resistant when faced with new situations, people or things? Slow-to-warm up children tend to think before they act. They are less likely to act impulsively during adolescence.

7. *Adaptability*: Related to how easily the child adapts to transitions and changes, like switching to a new activity. Does the child have difficulty with changes in routines, or with transitions from one activity to another? Does the child take a long time to become comfortable to new situations? A slow-to-adapt child is less likely to rush into dangerous situations, and may be less influenced by peer pressure.

8. *Persistence*: This is the length of time a child continues in activities in the face of obstacles. Does the child continue to work on a puzzle when he has difficulty with it or does he just move on to another activity? Is the child able to wait to have his needs met? Does the child react strongly when interrupted in an activity? When a child persists in an activity he is asked to stop, he is labeled as stubborn. When a child stays with a tough puzzle he is seen a being patient. The highly persistent child is more likely to succeed in reaching goals. A child with low persistence may develop strong social skills because he realizes other people can help.

9. *Mood:* This is the tendency to react to the world primarily in a positive or negative way. Does the child see the glass as half full? Does he focus on the positive aspects of life? Is the child generally in a happy mood? Or, does the child see the gall as half empty and tend to focus on the negative aspects of life? Is the child generally serious? Serious children tend to be analytical and evaluate situations carefully.

Temperament is the innate behavior style of an individual that seems to be biologically determined. Although some experts feel that labeling a child too quickly as "difficult" may create a self-fulfilling prophecy of problematic parent-child interaction, knowing what kind of temperament your child has may make the difference between a happy and a troubled child - and between an accepting and frustrated parent.

Coping Child's Personality

Ever feel frustrated by your high-energy baby? What can you do about a child who screams himself silly when he doesn't get his own way? A youngster who gets overexcited when a playmate come over? The experts tell us that there's probably not a lot you can go about changing the way a child tends to reach if that tendency is inborn but there are ways you can help him manage his impulses better - and spare yourself lots of grief along the way. Realize that your child's immature behavioral style is not your "fault" because temperament is biological not something he learned from you. Still it is within your power to help your child cope with his temperament - and eventually to understand himself better instead of feeling sorry for yourself for having a noisy, distractible or shy child. Learn to accept this as his nature and then develop a strategy to help him adapt in a socially acceptable way.

Replace a victimized mind-set with an adult resolve to help your child ameliorate his difficulties. Above all, remember that all temperamental qualities can be shaped to work to a child's advantage if they are sensibly managed. To become a "manager of your child's temperament, make sure that you step back from his objectionable behavior for a minute and remind yourself that his shrill shriek of excitement or his irregular sleeping habits are not deliberate reactions but one he can yet control. The key is to switch on the objective part of your mind rather than to become emotionally embroiled in his temperamental difficulties. Through this emotionally "neutral" stance, you'll be better able to help him modify his reactions because you'll be thinking rationally.

Develop specific plans ahead of time to cope with troublesome behavior and then enforce them in a sympathetic but consistently firm ways. If your child tend to get wild on family occasions or when he's with friends, be sensitive to this tendency and take steps to quiet it before it escalates. (Decide ahead if this activity is one he can handle. With younger children avoiding potential problem situations may be the best solution). With a baby you may want to tell your host that you will want to leave the party early. You can also take your child into a quiet room and sit with him until he falls asleep. Follow similar procedure with an older child, either by removing him from the activity, distracting him with something quieter such as a story hour or a snack or calling a "time out" period.

An infant with irregular biological rhythms will need special structuring from you so that he eventually learns to sleep through the night, to eat at the usual meal times and to control his bladder and bowel function. In this case, a doctor or child-behavior expert may be able to help to develop a schedule for your baby. For an older child who resists going to sleep, you may have to make special distinctions between bedtime and "sleep-time." To help him settle down, you can insist that the youngster get into bed at a certain time but permit him to read or play quietly until he feels sleepy. In this way, you are regulating his schedule but still allowing him to relax at his own pace.

Learn to distinguish between behavior that is temperamentally induced and that which is learned. If a child knocks over your best vase by mistake because he is a high-energy child and was running gleefully through the living room, your response should be different than if he broke your vase deliberately. In some instances you will probably be upset and may express your displeasure. But the action you pursue should be different. In the first case you may have to give some thought on how to prevent your child from running through the living room and remembering other ways he can work off his energy while in the house. In the second scenario, you will probably want to punish the child for his deliberate destruction of your personal property to impress upon him that this behavior is socially unacceptable. With temperament, the goal is always to manage rather than to systematically punish.

By the same token learn to distinguish between a tantrum that is temperamentally determined and one that is deliberately manipulative. Both may look the same because in both instances the child is crying or screaming loudly but the reasons for them are different. A strong-willed and intense child may react to a disappointment with a tantrum but the parent should understand that in a sense the child really can't help it - that this is his innate behavioral reaction. This is in marked contrast to the less intense child who screams and cries in the same way when you say no because he has learned that such behavior will weaken your resolve and make your give into him. Becoming an expert on your child's temperament will help you distinguish between the two types of tantrums - and then you can react to the tantrum appropriately.

Finally remember that one of the most important jobs a parent can do is help his child develop self-esteem. That doesn't mean over-inflating his ego but rather helping him develop a positive sense of himself with a fair sense of his strengths and weaknesses. Understanding a child temperament is the first step toward enhancing his self-esteem because you will be able to deliver praise sensitively in accordance with his innate tendencies and help him build upon those traits in a positive way.

Parenting strategies for very intense children

— Provide activities that are soothing such as warm bath, massage, water play, stories.

— Recognize cues that signal that intensity is rising.

— Help child learn to recognize cues that signal that intensity is rising.

— Use humor to diffuse intensity.

— Teach child to use time-out as a time to calm self-down.

— Avoid escalating intensity of child be reacting intensely to his/her behavior. Give calm, clear, brief feedback.

Parenting strategies for slow-to-adapt children

— Establish clear routines.

— Prepare child by discussing plans for the day when routine changes.

— Prepare child for transitions.

— Give warnings a few minutes before transition from one activity to next occurs.

— Allow time for closure of one activity before going on to next.

— Stay aware of number of transitions required, and keep transitions to minimum if possible.

ADOLESCENT STAGES OF DEVELOPMENT

Children must pass through several stages, or take specific steps, on their road to becoming adults. For most people, there are four or five such stages of growth where they learn certain things: infancy (birth to age two), early childhood (ages 3 to 8 years), later childhood (ages 9 to 12) and adolescence (ages 13 to 18). Persons 18 and over are considered adults in our society. Of course, there are some who will try to act older than their years. But, for the most part, most everybody grows in this same pattern. Parents learn much about taking care of their babies and young children. At the hospital or with the doctor, you might pick up information about what to feed them or how long they should sleep.

Later, school staff may remind you about the importance of talking and reading to your young children. You can also see how your friends or relatives treat their kids. You cannot say the same thing about learning to talk with teenagers (adolescents). It seems like everyone, even teachers and neighbors, have problems understanding them. Giving up, you might turn to doing and saying the same things your parents did with you. But those were other times! You can begin to understand this age group if you look at its place on the growth sequence. Notice how it's right next to the adult stage, the last step before being an adult. This is a time for adolescents to decide about their future line of work and think about starting their own families in a few years.

One of the first things they must do is to start making their own decisions. For example adolescents can begin to decide what to buy with their own money or who will be their friend. To do this they must put a little distance between themselves and their parents. This does not mean that you can't continue to "look after them" or help them when needed. You should, as much as possible, let them learn from the results of their actions. Adolescents also need to be around other adults, both male and female. These can be relatives, neighbors, or teachers. Of course, they should be positive role models. Your teenagers can learn from them about things like how to fix the car, getting along with others, or ideas for future jobs. Finally, don't worry if they want to spend time alone. Adolescents can "spend hours" day dreaming about their future life. They might be planning the things they can do or will buy "when they grow up." Remember, to travel far, one begins with the first few steps!

ADOLESCENCE: THE LAST STEP BEFORE BECOMING AN ADULT

Adolescents face a range of developmental issues. Havighurst suggested that two important areas included work and relationships. Levinson focused on changing relationships and on exploration, while Erikson commented on intimacy and commitment to goals. Super indicated that exploring and crystallizing vocational choice are important to older adolescents and young adults. What seems evident is that older adolescents and young adults enter transitions with the goal of becoming independently functioning adults, as they strive to meet evolving personal and career related needs. Rapid and escalating changes in labor market and post-secondary educational opportunities mean that adolescents now are confronted with the challenge of meeting their personal and career needs when neither can offer certainty or a sense of personal control.

Transition From High School

A longitudinal study by Amundson, Borgen, and Tench (in press) found that young people left high school unprepared for current career realities and that both the career and personal areas of their lives were in a state of change and uncertainty. At the end of their final year of high school, young people in the study expressed optimism about entering the career area of their choice and they expected to be successful workers in challenging jobs which offered personal satisfaction. About half the respondents indicated some concern about meeting post-secondary entrance standards. Approximately 9 and 18 months following graduation, depression, self-esteem, and anxiety were correlated with a range of perceived problems, including money, lack of support from family and friends, internal attribution of general transition problems, external attribution of career/ employment difficulties, and lack of job satisfaction.

At the end of the study, some of the young people were interviewed. They were asked about factors that helped or hindered the post-high-school transition. Positive factors included supportive family and friends, making money, satisfying leisure activities, personal achievements, and educational success. Negative factors included relationship problems, career confusion, financial difficulties, unemployment, lack of satisfying work, lack of post-secondary educational opportunities, and difficulty in adjusting to post-secondary educational demands. Developmentally, the young people were trying to meet personal and career-related needs, which were in a state of flux and uncertainty. It was apparent that a lack of progress in one area could have a negative influence on the other (e.g., an inability to gain post-secondary educational admission or paid work could drastically alter one's ability to move from being a dependent adolescent to an independent adult).

The above study suggests a need for a broader view of career counseling; counseling which recognizes the developmental needs of young people, the influence of social and economic changes, and the importance of basing intervention strategies on personal and career competence, all within a context of diminished and changing opportunities for choice. In order to address this broader range of issues, there are competence model with eight main areas: purpose, problem solving, communication skills, theoretical knowledge, applied knowledge, organizational adaptability, human-relations skills, and self-confidence. Also have developed a number of counseling strategies that facilitate a smoother transition:

— *Developing Multiple Plans*: Many young people leave high school with a narrow plan of action and with few alternatives. They fully expect to be successful with the plan and are not prepared to face any barriers. Developing flexibility in career planning requires a sense of purpose, problem solving skills, and several plans. Helpful strategies include visualization, lateral thinking, assessing options, and decision making in a context of uncertainty.

— *Self-Advocacy and Marketing*: As young people move towards further education, or into the labor market, it is critical for them to market and advocate for themselves. With scarce opportunities and confusing bureaucracies, there is a need to develop communication skills, self-confidence, organizational adaptability, and effectiveness in human relations. This requires activities such as mentoring, role-played practice, and ongoing economic, emotional, and informational support.

— *Managing Changing Relationships*: The emotional and social changes adolescents experience can challenge young people as they try to cope with barriers in the education system and labor market. Friends provide emotional support, but this is a time when friendship patterns are changing. Parents are needed for emotional, material, and information support, but, at the same time, they need to allow young people sufficient room to develop their own sense of identity. Coping with relationship issues can be facilitated through communication, human relationship training, and problem solving, which blurs most of the traditional distinction between career and personal counseling.

— *Meeting Basic Needs*: Young people have a strong need for community. Other central needs include having a sense of meaning in life, physical and emotional security, and basic structure in relationships and living. As young people mover beyond high school, many of these basic needs require revaluation. In addition to changing relationships, questions emerge as to how to make a living, how to plan meaningful activities, and how to effectively manage time. To facilitate these changes, young people need to establish a sense of purpose and understand how they are meeting their current and future needs. Counselors can help clarify these issues. Without this type of developmental assistance, young people often lack the resilience to maneuver within increasingly competitive educational and labor market environments.

— *Coping with Stress*: Adolescence is a period of considerable stress. While much of the stress can by minimized through support, persistence, and active decision making and planning, there still will be times when young people find themselves in difficult situations. Coping with stress is associated with various competencies such as organizational adaptability, human relations, problem solving, and self-confidence. Particular strategies for stress management include relaxation techniques, managing 'self talk,' focusing, and using support systems.

— *Coping with Loss*: Young people were influenced by various personal losses. These losses involved death in the family (usually grand parents) and the experience of parental separation and divorce. The impact of these losses upon career events was considerable, suggesting a definite need for youth to develop competence in handling loss and grieving. Counseling in this domain blurs many of the traditional distinctions between the personal and career areas.

— *Bridging Programs*: Many young people lack "hands-on" experience as they attempt to enter the world of work. Many also are unfamiliar with, and fearful of, moving into post-secondary education. To address this concern, counselors need to develop work experience and co-op education programs to help young people acquire the necessary experience. Post-secondary education entry programs can also play an important role in easing transition difficulties.

— *Information and Information Access*: The challenge in the information age is not only how to gather information, but how to turn information into personally relevant knowledge. Young people need up-to-date information on careers, education programs, and market trends. They must also develop skills to assess the relevance of information. Acquiring these skills involves both theoretical and applied knowledge. Counseling strategies within this domain include helping young people develop research, interviewing, and critical analysis skills.

GOOD NUTRITION FOR KIDS & TEENS

All parents want their children to be healthy. As elementary school aged children go through remarkable physical changes of all kinds, their food intake becomes a critical aspect of this growth and development. Recent research shows that nourishing food not only makes a child healthier, it makes him emotionally more stable, and it improves school performance. It appears then that paying attention to our children's diets pays high dividends. If only our children thought so, too! Because children tend to rank their parents' views on food along with their unpopular views on curfews, rock music, hair styles, etc., it is up to the parents to, first of all, be clever about insinuating nutritious foods into the family menus and, secondly, take a reasonable but hard line when other approaches fail.

Breakfast

A child in the classroom whose last meat was dinner the night before has gone about sixteen hours without food, and that child is hungry, whether he knows it or not. A nutritious breakfast will provide energy for several hours-until lunch, in fact. Is any kind of breakfast better than no breakfast at all? Unfortunately, no. A doughnut, for example, provides a quick rush of energy that lasts about 40 minutes, about the length of time it takes the youngster to get from the breakfast table to his classroom! Traditionally, teachers schedule "heavy" subjects, such as reading and arithmetic, during the morning hours, and so it becomes even more important that the child's brain be fueled. The following suggestions have proved helpful in sending youngsters off to school ready to team.

— Put the blender to good use. Concoct a shake or smoothie with milk, vanilla, and a couple of tablespoons of honey. There is an unlimited assortment of blended breakfast drinks with all sorts of combinations: orange juice, bananas, apple juice, wheat germ, etc. Eight ounces of such a drink served over shaved ice with a couple of slices of whole, wheat toast will keep any youngster on his toes until noon.

— If the youngster likes cereal, dry or cooked, give it an extra boost with a sprinkling of wheat germ. chopped nuts, raisins, or other fruit.

— Leftover pizza doesn't make a bad breakfast. It's more nutritious than any other fast food, and you can increase its nutrition by adding extra cheese.

— Layer yogurt, fruit, and granola in glasses for a parfait look.

— Try a breakfast buffet with sliced fresh fruits, finger food vegetables, hard-boiled eggs, whole-wheat muffins. The more colorful, the better, and kids love to help themselves.

Lunch

The sack lunch! Does the child give, trade, or throw most of it away? (The clue is if he comes home ready to eat anything and everything in the refrigerator!) It's altogether possible that the youngster is jettisoning his sack lunch because he is bored to tears with it, so it's time for the parents to get creative. One clever parent inserts a smaller bag labeled "this is for trading" into the larger bag, and it seems to work wonders! The following suggestions may help.

— Apples and oranges certainly qualify on all counts-nutritious, no preparation, relatively inexpensive. But they're also easy to toss in the garbage can! Get a couple of plastic containers with lids that stay on and fill them with fruit cocktail, applesauce, mandarin orange slices, yogurt, even popcorn or Crackerjacks.

— Use cookie cutters to shape sandwiches, crinkle-cut carrots, stuff some celery. If your youngster is a peanut butter addict-and most are-add any of the following to peanut butter for a change from the tried and true peanut butter and jelly sandwich: chopped dates or nuts, raisins, bacon bits, applesauce, crushed pineapple.

— Mix tuna fish or canned salmon with sliced cucumbers, sprouts, grated carrots hard boiled eggs, chopped celery, etc.

After School Snacks

Once a child has entered the primary grades, it is no longer possible for him to eat whenever he feels hungry, and it can be a long time between lunch and the final bell. Most youngsters arrive home wanting and needing an immediate energy boost. It's a great opportunity to add some "in nutrients to the youngster's diet. To many youngsters a snack automatically means something sweet; however, sugar should be removed from the diet as much as possible except for special occasions.

— Post a "what's inside for a snack" list on the refrigerator door and let the child help himself.

— Select a special spot where the child will find either that day's snack or a note telling him where to find it.

— Keep a supply of trail mix in a moisture-proof container. The combinations are limitless: nuts, coconut shavings, dates, sunflower seeds, pretzel sticks, banana chips, etc.

— Most youngsters like dried fruits, especially if they can spread them with peanut butter or a fruit butter.

— Popcorn-youngsters love it as a snack, and it can also be sprinkled on soup.

— Cheese, cheese, cheese-spread, melted or in chunks. Whole wheat crackers.

Dinner

The time-honored tradition of breakfast, lunch, and dinner seems the best way to ensure a balanced diet, but the fact is that we may not need three meals a day. A better solution for some families may be more frequent, lighter meals. And it really isn't a matter of life and death if a family member misses dinner. No child ever starved to death because he was playing softball and forgot to come home for dinner. Common sense, flexibility, and creativity go a long way to make the evening meat pleasant.

— Let the youngsters serve themselves In this way they can decide how much to put on their plates and can always take a second helping if they want it.

— Children should be expected to taste every dish that's been prepared. If they don't like it, they don't have to eat any more but it's an excellent way to expand their food horizons.

— Make food look attractive and interesting. One mother cuts liver into bite-sized pieces and sticks toothpicks in each piece. Her children eat the pieces lollipop-style with a great degree of gusto!

— Involve your youngsters in the family menu. Let them suggest foods, familiar and unfamiliar, although some of their choices may need to be discussed in terms of whether or not they fit into the family budget.

— Occasionally take your children to the grocery store. (Be sure to feed them and yourself before you go or your food bill will soar!) It's a good place to give them choices. ("Do you want broiled chicken or fish for dinner? Peas or green beans? " "Pick out some fruit for your lunches this week") What you don't buy is as important as what you do buy. If you don't buy potato chips and sodas, your youngsters won't be able to snack on them.

— Turn off the television set. Make the dinner table a place for good conversation.

Most of the foods served in fast food restaurants have fat as their main source of calories. Even milk shakes are often nude with highly saturated coconut oil. In addition, their foods are usually low in iron, fiber, and vitamins, and extremely high in sodium. Unfortunately, youngsters are exposed to virtually thousands of junk food television commercials a year, and parents might as well accept the fact that occasionally their children are going to head for a fast food restaurant. However, they'll survive, especially if their daily diet is nutritious.

CHILD'S PHYSICAL DEVELOPMENT

Babies and young children learn primarily through movement and their senses how to deal with gravity, to keep their balance, to move their body through space, about time and sequence of events. A child's growth is a continuous process, a gradual sequencing from one stage of physical and mental development to another-"Each child sits before he stands; he babbles before he talks" (Gesell). It's a marvelous process to watch and a marvelous opportunity for parents to foster and implement important periods of growth.

In reality, the responsibility of the parent is twofold. Not only should proper emphasis be placed on movement and exercise in the home but whenever possible the school's approach to physical education should be monitored. A good P.E. program in elementary school ideally will offer three or four periods a week of 45-60 minutes' duration. The program need not necessarily be highly structured and should certainly not be highly competitive. Movement is the key, and that can include simple activities (running in place, jumping jacks) and games. The P.E. program should progress developmentally from grade to grade and should be designed to offer maximum benefit to every child, no matter how small or Late-maturing.

Parents must be careful of physical conditions that might limit a child's movements and participation. Most schools ask, for their records, that a medical report be on file at the school, but it is the parent's responsibility to see that the report is accurate and up to date and that everyone on the faculty who needs to be aware of the report knows about it.

For youngsters who are interested in competitive sports, almost every community offers after-school and summer sports such as soccer, baseball, and football. But *these highly* organized activities can promote stress if emphasis is placed on winning rather than just enjoying the game. An observant parent can usually quickly tell if the child is paying a high emotional price rather than just having fun. And it should be noted that in some highly organized sports, the youngsters spend more time standing around and watching than actually participating. The local YMCA and YWCA usually offer well-rounded pro- grams that include fitness exercises and swimming The fitness program may consist of carefully structured aerobics, and the swimming program is usually designed for individual mastery rather than competition.

Parents are enormously busy people-perhaps both parents work outside the family; perhaps there are several children in the family with differing needs and demands; perhaps it is a one-parent family. The activities that follow are offered with precisely those situations in mind. They are simple, inexpensive, enjoyable, arid can be adapted for groups (the whole family and/or friends) as well as for individual youngsters.

Simple Motor Activities

Keep a simple record of your child's physical development. Every year on his birthday, write down his weight and height.

Find a convenient wall space, place a ruler on the child's head, draw a line, and date it. Children love to watch how much they have grown. While your child is standing in place, have him count the number of times he can go up and down on his toes. Set aside time in the family schedule for a family walk, perhaps just 15 minutes, or a Saturday afternoon leisurely hike for an hour or more depending on the youngster's age and stamina.

A family walk is a great way for parents and siblings to interact and chat-something that is often difficult to fit into the busy lifestyles of the nuclear family. Walks can also provide an in-depth look at changes in Mother Nature and the community during different seasons of the year. And then there are the very simple motor activities: hopping, jumping, skipping, and climbing. All are important in a child's growth patterns. Each one calls upon various muscle groups to require extensive use. Remember hopscotch? All that is needed is a piece of chalk and a couple of pebbles. If parents will recall their own childhood, they may tap into some games that were fun and that, without knowing it; build strong bones and muscles. Try rolling-on a level plane or on a hill. How many different ways can the child roll? Arms outstretched; arms at sides; one arm stretched the other to the side.

Head and neck exercises: Turn head side to side, down and up, while standing, sitting, lying on the back and on the stomach. Have the child walk across a fallen log or along a narrow curb. Have him repeat the walk, holding a bulky object in one hand, then the other hand, over his head. Repeat going backwa rd and sideways.

Row a boat on dry land: The child must calculate which oat to use in order to turn a specific direction. (The parent will have to figure this out first!) Water activities for pool, lake, or rub (be prepared to get splashed!). Hold a ball and ask the child to hit it with his hands (right and left), elbows, knees, feet. If swimming lessons are available, enroll your youngster. The earlier, the better. Just tossing a ball from parent to child is excellent for eye-hand coordination as well as large muscles. Don't let the activity get boring. Vary it by asking the child to kick the ball (using alternate feet) or batting it (with alternate hands). Ball size is important. Large enough for a success experience. Small enough for a challenging experience. Don't forget beanbags-quite a different experience from throwing or catching a ball. Let the child toss and catch it himself-standing, sitting, lying down, alternate hands. Can he catch it on the top of his hand? a shoulder? a knee? a foot?

Differing chairs: The child sits down and gets up from chairs and stools of varying heights, descending and standing up slowly and without using his hands. The lower the chair, the more difficult the task.

— *Kangaroo hop*: Have the child hold something (for example, a beanbag-or if you want to make it difficult, an apple or an orange) between his knees, then jump with feet together. Frontwards, backwards, sideways.

— *Save your large bleach bottles*: With the bottoms cut off, they make nice scoops for catching games, using tight objects such as a whiffle ball or beanbag.

— *Wheelbarrow*: Hold the child's legs white he "walks', with his hands along a marked route.

Find a place where the child can see his shadow. Then see how creative you can be in directing his activities: "Make your shadow tall, short, wide, thin, make it jump, stand on one foot, touch its feet," etc. Most of the activities that have been described can, for the most part, be done inside or out. It is important that they be done in a spirit of good fun and recreation. Once they become a chore, the child, either subtly or overtly, will decrease his effort and the sought after physical development will diminish. The secret probably ties in offering a variety of activities with an attitude of good cheer. And there may be a bonus - parents may discover that they, too, are in better shape!

PHYSICAL FITNESS GUIDE FOR KIDS AND TEENS

Fitness and 2-3 Year Old

Fitness that starts in childhood is a lifelong investment. Studies have shown that children who participate in fitness activities continue to stay active throughout their lives. Developing the habit of fitness promotes healthful behaviors, improves self-esteem, and decreases the risk of serious illnesses such as heart disease and stroke later in life. In addition, regular physical activity at any age helps your child to play and meet challenges. Part of committing your child to fitness may include limiting your own excessive computer or television time, as well as becoming a role model by getting regular exercise on your own or with your children. Remember that you are your child's best role model, so always pay attention to safety in sports and recreation.

Chronic health conditions

Children with chronic health conditions and disabilities should not be excluded from fitness activities; they receive the same positive benefits from exercise. Some activities may need to be modified or adapted to your child's disability. Certain activities are dangerous for some health conditions. Consult your child's doctor about the safety of fitness activities for your child with a disability. What exercises can kids in this age group do to stay fit? Toddlers use play to explore and learn about the world. By encouraging even the youngest child to engage in active play, parents prompt a child's exploration and natural curiosity about how his body works.

Two- and 3-year-olds thrive on unstructured play. To help toddlers develop skills, running, swinging, climbing, playing in a sandbox, and carefully supervised waterplay are all fun and healthy. By age 2, active, fit children should be able to jump with two feet, skip and run. By age 3, they should be able to change directions (from left to right, from forward to backward) comfortably. Although there are a number of programs to teach toddlers how to swim, do gymnastics, or even begin to learn baseball basics in peewee leagues, parents should carefully investigate these programs, with an eye toward safety, philosophy, and age-appropriateness. Most children this age are not developmentally ready to play in organized sports or to be involved in competitive activities. If your child doesn't want to participate in an activity for any reason, don't force your child.

Try to find out if there are any fears or reasons why your child is reluctant to join in the play. If necessary, postpone the activity, and introduce it later. Waiting a few years and then suggesting the activity again is probably the wisest course. If your toddler refuses to play or interact with peers, it can be an indication of a physical or psychological problem. Children who complain of pain when they play or consistently refuse to join other children in outdoor play may need to be seen by a health care professional. Even a shy child needs to play with other kids. Socialization skills are important to future functioning. You may want to seek out child play groups to give your child more opportunities for peer interaction.

Family Fitness Tips

Any number of activities can be incorporated into family fitness. Walking, playing and running in the backyard or using playground equipment at a local park can be fun for the entire family. Always keep safety in mind: small fingers can get easily caught in playground equipment, and wheeled objects such as tricycles or bicycles can promote sudden twists or falls. When engaging in family fitness outings, remember these tips:

— All children and adults should always wear helmets when on tricycles, bicycles and big wheels; wearing a helmet will become a lifelong habit if started early.

— Monitor children when they cross streets; even if your child has been instructed to "look both ways," kids engrossed in a neighborhood game can easily dash after a ball without thinking.

— Safety is a particular concern when it comes to water sports. Never let your toddler around a pool without your constant supervision. It takes very little time for a toddler to slip beneath the water. Swimming programs for young children have an eye on safety and teach valuable skills.

— Use sunscreen whenever your child is playing outdoors, even on overcast days, to prevent sunburn and decrease the risk of developing skin cancer at a later age.

— One common activity that is not recommended is the use of trampolines. The American Academy of Pediatrics recommends that all trampolines, including home trampolines, be avoided due to the high number of injuries at all ages.

Fitness and 4-5 Year Old

Fitness that begins in childhood is a lifelong investment. Studies have shown that children who participate in fitness activities continue to stay active throughout their lives. In addition, regular physical activity helps your child to play and meet challenges, and it's perfect for children of all ages. Part of committing your child to fitness is limiting your own excessive computer or television time, as well as becoming a role model by getting regular exercise on your own or with your children. Remember that you are your child's best role model, so always pay attention to safety in sports and recreation.

Chronic Health Conditions

Children with chronic health conditions and disabilities should not be excluded from fitness activities; they receive the same positive benefits from exercise. Some activities may need to be modified or adapted to your child's disability. Certain activities are dangerous for some health conditions. Consult your child's doctor about the safety of fitness activities for your child with a disability. What exercises can kids in this age group do to stay fit?

As children develop, so do their physical skills. Children ages 4 and 5 play in an increasingly coordinated manner and can participate in organized games. Four- and 5-year-olds can roll large balls, play catch and may be able to navigate a bike with training wheels. However, children this age cannot safely maneuver a bicycle in areas where there is traffic because they lack judgment and safety awareness as well as coordination skills. Swimming is an excellent fitness activity and teaches a valuable skill. Other fitness activities they may enjoy include dance, skiing or skating.

No matter what the sport or activity, parents should remember that events should always be fun. If your child isn't having fun, ask why and try to resolve what is bothering your child or find some other fitness activity. This is particularly true of organized sports. Kids who are pressured to compete may develop a negative attitude toward fitness or injure themselves while trying to please others with their performance. Make sure your child's time is not overscheduled; this may cause stress. Unstructured time allows your child to learn important skills and time to wind down.

If your child refuses to participate in any fitness activity, it can be an indication of a physical or psychological problem. Children who complain of pain when they play or consistently refuse to join other children in outdoor play may need to be seen by a doctor. Even a shy child needs to play with other kids.

Family Fitness Tips

Any number of activities can be incorporated into family fitness. Walking, playing and running in the backyard or using playground equipment at a local park can be fun for the entire family. Sports such as tennis, skiing, dancing, ice-skating and hiking provide opportunities for family fun, as well as offering children a chance to build skills in sports that can bring them pleasure for a lifetime. The biggest concern with children this age is safety. Small fingers can get easily caught in playground equipment, and wheeled objects (such as tricycles, bicycles or in-line skates) can promote sudden twists or falls. When engaging in family fitness outings, keep these tips in mind:

— All children must wear helmets when on wheeled vehicles. Establishing this habit early is vital.

— Watch while children cross streets; even if your child has been instructed to "look both ways," kids engrossed in a neighborhood game can easily dash after a ball without thinking.

— Safety is a particular concern when it comes to water sports. It takes very little time for a child to slip beneath the water.

— Always apply sunscreen when playing outdoors, even on overcast days.

— One common activity that is not recommended is the use of trampolines. The American Academy of Pediatrics recommends that all trampolines, including home trampolines, be avoided due to the high number of injuries at all ages.

Fitness 6-12 Year Old

Fitness that begins in childhood is a lifelong investment. The regular habit of exercise is important to develop during this stage to prevent obesity. Kids should learn to develop exercise as a lifelong habit. Studies have shown that children who participate in fitness activities continue to stay active throughout their lives. In addition, regular physical activity helps your child to play and meet challenges, and it's perfect for children of all ages. Part of committing your child to fitness is limiting your own excessive computer or television time, as well as becoming a role model by getting regular exercise on your own or with your children. Moms or dads can show children the importance of sports activities by playing along, coaching teams, or simply coming out to support their child's games. Remember that you are your child's best role model, so always pay attention to safety in sports and recreation.

Chronic Health Conditions

Children with chronic health conditions and disabilities should not be excluded from fitness activities; they receive the same positive benefits from exercise. Some activities may need to be modified or adapted to your child's disability. Certain activities are dangerous for some health conditions. Consult your child's doctor about the safety of fitness activities for your child with a disability. What exercises can kids in this age group do to stay fit?

As children develop, so do their abilities to participate in a variety of sporting activities. Elementary-school-aged children, particularly those between the ages of 10 and 12, are poised for new learning experiences and have the ability to absorb new skills required for both team and individual sports quickly. Even children who prefer not to participate in organized teams need regular exercise. Any number of noncompetitive sports, such as swimming, ice-skating, gymnastics, dance and nonviolent martial arts, can help keep your child healthy and fit. At age 11, children who are interested may begin to lift light weights (under supervision) to build muscle to help prevent future sports injuries.

What should parents do if they're concerned about their child's fitness? If your child refuses to participate in any fitness activity, it can be an indication of a physical or psychological problem. Children who complain of pain when they play or consistently refuse to join other children in outdoor play may need to be seen by a doctor. Even a shy child needs to play with other kids. Once children begin to participate in sports, injuries may occur. Many of these arise from the overuse of certain muscles, or because children's skeletal and muscular systems are not yet fully developed. These growing muscles, ligaments and tendons are vulnerable to injury. During puberty, which can begin as early as 8 1/2 years in girls and 9 years in boys, children experience a growth spurt, when bones grow more quickly than muscles and tendons, making muscles and tendons short, tight and easily injured.

In addition, preteens going through puberty may also become less coordinated as they adjust to their physical changes, which can increase the risk of injury. Although many injuries will respond to RICE (Rest, Ice, Compression and Elevation) if your child seems to be in severe pain or the injury doesn't seem to be getting any better, you should see your child's doctor. Appropriate warm-ups and stretching decrease the chance of muscle and tendon injury. Young athletes, particularly those in gymnastics, wrestling or dance, may develop eating disorders. If your son or daughter refuses to eat certain food groups (such as fats), becomes overly concerned with their body image, or experiences a sudden loss or gain in weight, you should talk to your child about your concerns. If your child doesn't respond to your discussion, consult your child's doctor. Early intervention is vital.

Family Fitness Tips

Any number of activities can be incorporated into a family fitness activity. Walking, bike riding, camping and hiking provide opportunities for fitness and fun. Tennis, skiing, dancing and ice-skating offer children a chance to build skills in sports that can bring them pleasure for a lifetime. Whether children are participating in sports or individual fitness activities, safety should be paramount. When engaging in family fitness outings, keep these tips in mind:

— Parents should be sure that kids always wear safety helmets when on wheeled vehicles.

— Young athletes should be taught the importance of stretching their growing muscles before any strenuous physical activity to prevent tears and strains.

— Your child's coach or trainer should be familiar with how to deal with injuries such as concussions. Children who have had even a mild concussion and who did not lose consciousness may only be confused, but may still be at risk of serious injury if they have another concussion before complete recovery.

— Always apply sunscreen when playing outdoors, even on overcast days.

— One common activity that is not recommended is the use of trampolines. The American Academy of Pediatrics recommends that all trampolines, including home trampolines, be avoided due to the high number of injuries at all ages.

Fitness 13-18 Year Old

Fitness that begins in childhood is a lifelong investment. Studies have shown that kids who participate in fitness activities continue to stay active throughout their lives. In addition, regular physical activity helps your child to play and meet challenges, and it's perfect for children of all ages. But research has also shown that from the ages of 13 to 18, there is a steady downward trend in fitness levels. Part of this may be due to the number of other activities available to teens, along with personal choices to not participate in sports. Remember that you are your child's best role model, so always pay attention to safety in sports and recreation.

Chronic Health Conditions

Children with chronic health conditions and disabilities should not be excluded from fitness activities; they receive the same positive benefits from exercise. Some activities may need to be modified or adapted to your child's disability. Certain activities are dangerous for some health conditions. Consult your child's doctor about the safety of fitness activities for your child with a disability. Teenagers need to exercise to avoid high cholesterol, obesity and other health risks. Beyond the physical benefits, sports can provide teens with self-confidence, a group of like-minded friends, and a chance to excel at something they love. Such benefits are not limited to competitive sports; noncompetitive activities can help teens test themselves against their own limits.

What exercises can kids in this age group do to stay fit? Developmentally, teenagers can pick up on any activity that they enjoy, from competitive to noncompetitive sports. In addition to focused competitive sports, skateboarding, in-line skating, rock climbing or snowboarding all qualify as great fitness activities. Weight training, under supervision, can build muscle to help prevent other sports injuries. Although safety (including the proper equipment) is an important priority, any activity can be a good source of fitness.

Like adults, teens should aim to exercise at least three times a week for 30 minutes at a time and pursue an active lifestyle to help prevent future health problems. What should parents do if they're concerned about their child's fitness? If your child refuses to participate in any fitness activity, it can be an indication of a physical or psychological problem. Children who complain of pain when they play or consistently refuse to join other children in outdoor play may need to be seen by a doctor. Even a shy child needs to play with other kids.

When children begin to participate in sports, injuries may occur. Many of these arise from the overuse of certain muscles, or because children's skeletal and muscular systems are not yet fully developed. These growing muscles, ligaments and tendons are vulnerable to injury. During puberty, which can begin as early as 81/2 years in girls and 9 years in boys, children experience a growth spurt, when bones grow more quickly than muscles and tendons, making muscles and tendons short, tight and easily injured. In addition, preteens going through puberty may also become less coordinated as they adjust to their physical changes, which can increase the risk of injury. Although injuries will respond to RICE (Rest, Ice, Compression and Elevation) if the child seems to be in severe pain or the injury doesn't seem to be getting any better, you should see your child's doctor. Appropriate warm-ups and stretching decrease the chance of muscle and tendon injury.

Young athletes, particularly those in gymnastics, wrestling or dance, may develop eating disorders. If your son or daughter refuses to eat certain food groups (such as fats), becomes overly concerned with his or her body image, or experiences a sudden loss or gain in weight, you should talk to your child about the problem. If your child does not seem to improve, consult your child's doctor. Early intervention is vital. Another issue that can rise at this time is the use of steroids. There is no reason for any teenager to use them. If you suspect your child is taking them, consult your child's doctor at once.

Family Fitness Tips

Any number of activities can be incorporated into a family fitness activity. Walking, bike riding, camping and hiking provide chances for fitness and fun. Tennis, skiing, dancing and ice-skating offer teens a chance to build competent skills in sports that can bring them pleasure for a lifetime. Whether kids are participating in sports or individual fitness activities, safety should be paramount. When engaging in family fitness outings, keep these tips in mind:

— Parents should be sure that teenagers continue to wear safety helmets when on wheeled vehicles.

— Young athletes should be taught the importance of stretching their growing muscles before any strenuous physical activity to prevent tears and strains.

— Your child's coach or trainer should be familiar with how to deal with injuries such as concussions. Children who have had even a mild concussion and who did not lose consciousness may only be confused, but may still be at risk of serious injury if they have another concussion before complete recovery.

— Always apply sunscreen when playing outdoors, even on overcast days.

— One common activity that is not recommended is the use of trampolines. The American Academy of Pediatrics recommends that all trampolines, including home trampolines, be avoided due to the high number of injuries at all ages.

Physical Fitness, Sports and the School-Age Child

As a parent, you need to encourage healthy habits - including exercise - in your youngsters. Physical activity should become as routine a part of their lives as eating and sleeping. Reassure them that sports such as cycling (always with a helmet), swimming, basketball, jogging, walking briskly, cross-country skiing, dancing, aerobics and soccer, played regularly, are not only fun but can promote health. Some sports, like baseball, that require only sporadic activity are beneficial in a number of ways, but they do not promote fitness. Physical activity can be healthful in the following ways:

— Increase cardiovascular endurance

— Improve large muscle strength and endurance

— Increase flexibility

— Maintain proper weight

— Reduce stress

— Increase Cardiovascular Endurance

More Americans die from heart disease than any other ailment; regular physical activity can help protect against heart problems. Exercise can improve your child's fitness, make him feel better and strengthen his cardiovascular system. Aerobic activity can make the heart pump more efficiently, thus reducing the incidence of high blood pressure. It can also raise blood levels of HDL (high-density lipoprotein) cholesterol, the "good" form of cholesterol that removes excess fats from the bloodstream. Even though most cardiovascular diseases are thought to be illnesses of adulthood, fatty deposits have been detected in the arteries of children as young as age 3, and high blood pressure exists in about 5 percent of youngsters.

At least three times a week, your middle-years child needs to exercise continuously for 20 minutes to 30 minutes at a heart rate above his resting level. As a guideline, the effort involved in continuous brisk walking is adequate to maintain fitness. Each exercise session should be preceded and followed by a gradual warm-up and cool-down period, allowing muscles, joints and the cardiovascular system to ease into and out of vigorous activity, thus helping to guarantee a safe workout. This can be accomplished by stretching for a few minutes before and after exercise.

Improve large muscle strength and endurance

As your child's muscles become stronger, he will be able to exercise for longer periods of time, as well as protect himself from injuries - strong muscles provide better support for the joints. Modified sit-ups (knees bent, feet on the ground) can build up abdominal muscles, increase lung capacity and protect against back injuries. For upper body strength, he can perform modified pull-ups (keeping the arms flexed while hanging from a horizontal bar) and modified push-ups (positioning the knees on the ground while extending the arms at the elbow).

Increase flexibility

For complete physical fitness, children need to be able to twist and bend their bodies through the full range of normal motions without overexerting themselves or causing injury. When children are flexible like this, they are more agile. Although most people lose flexibility as they age, this process can be retarded by stretching to maintain suppleness throughout life, beginning in childhood. Stretching exercises are the best way to maintain or improve flexibility, and they can be incorporated into your child's warm-up and cool-down routines. In most stretching exercises, your child should stretch to a position where he begins to feel tightness but not pain, then hold steady for 20 seconds to 30 seconds before relaxing. He should not bounce as he stretches, since this can cause injury to the muscles or tendons.

Maintain proper weight

Twelve percent of children in the prepuberty years are overweight, but few of these youngsters are physically active. Exercise can effectively burn calories and fat and reduce appetite. Ask your pediatrician to help you determine whether your youngster has a healthy percentage of body fat for his or her age and sex.

Reduce stress

Unmanaged stress can cause muscle tightness, which can contribute to headaches, stomachaches and other types of discomfort. Your child needs to learn not only to recognize stress in his body but also to diffuse it effectively. Exercise is one of the best ways to control stress. A physically active child is less likely to experience stress-related symptoms than his more sedentary peers.

Perhaps the best way to get children to enjoy exercise is to make it a family affair, with parents setting a good example and encouraging the youngsters to join the fun. The entire family can participate together in many physical activities, from swimming to cycling to hiking. Not only will everyone's fitness improve, but the unity of the family can be strengthened too. If you can get your child interested in fitness at a young age, you will improve the chances that physical activity will become a lifetime habit. Long-term health will be the ultimate result.

PUBERTY DEVELOPMENT

Puberty Information for Parents and Kids

As children approach and move through adolescence, they proceed through the following five stages of puberty:

Boys Girls

1. Stage one Prepubertal
   - No sexual development Prepubertal
   - No sexual development
2. Stage two Testes enlarge
   - Body odor Breast budding
   - First pubic hair
   - Body odor
   - Height spurt
3. Stage three Penis enlarges
   - Pubic hair starts growing
   - Ejaculation (wet dreams) Breasts enlarge
   - Pubic hair darkens, becomes curlier
   - Vaginal discharge
4. Stage four Continued enlargement of testes and penis
   - Penis and scrotal sac deepen in color
   - Pubic hair curlier and coarser
   - Height spurt
   - Male breast development Onset of menstruation
   - Nipple is distinct from areola
   - Stage five Fully mature adult

— Pubic hair extends to inner thighs

— Increases in height slow, then stop Fully mature female

— Pubic hair extends to inner thighs

— Increases in height slow, then stop

Children go through tremendous physical changes during puberty. The following guidelines can help parents along with children and young teens understand and cope with these changes:

Pubertal Development

Puberty often begins earlier than parents think. Breast budding in girls - their first sign of puberty - starts at age 10 on average, with some girls starting as early as 8 and others not starting until 13. The peak growth period (in height, weight, muscle mass and the like) in girls occurs about one year after puberty has begun. Menstruation usually starts about two years after the onset of puberty; on average, the first menses occurs just before girls turn 13.

Boys enter puberty about one year later than girls. The first sign is enlargement of the testes and a thinning and reddening of the scrotum, which happens at an average age of 11 but may occur anytime between 9 to 14 years. For boys, the peak growth period occurs about two years after the beginning of puberty. Puberty is made up of a clear sequence of stages, affecting the skeletal, muscular, reproductive and nearly all other bodily systems. Although boys and girls are generally of similar height during middle childhood, that changes with the beginning of puberty. Particularly in junior high school, girls are often taller than their male classmates, but within a year or two, boys catch up and usually surpass their female classmates. About 25 percent of human growth in height occurs during puberty.

There are many opportunities during this time of life for you to talk to your child about what she's experiencing. Your child needs to understand the physical changes that will occur in her body during puberty. You should emphasize that these changes are part of the natural process of growing into adulthood, stimulated by hormones (chemicals that are produced within the body). Also, while fully respecting her desire for privacy, keep track of your child's bodily changes. As the age ranges above indicate, there are wide variations of "normal" in the time when puberty begins; remind your youngster that while she and her friends will grow at different rates, they will eventually catch up with one another. Avoid even good-natured teasing of your child about his or her pubertal development. Because most youngsters feel self-conscious during this time, they will become embarrassed if they are kidded about the changing shape of their bodies or their deepening voices.

On occasion, youngsters start puberty either very early or very late. There is no need to overreact to this phenomenon. Even so, girls should be checked by their physician if they begin pubertal changes before age 8, while boys should be evaluated if they enter puberty before age 9. Likewise, see a doctor if there are no pubertal changes in a girl by age 13 or a boy by age 14.

Stages of Puberty

As children approach and move through adolescence, they proceed through the following five stages of puberty:

Boys Girls

1. Stage one Prepubertal

   — No sexual development Prepubertal

   — No sexual development

2. Stage two Testes enlarge

   — Body odor Breast budding

   — First pubic hair

— Body odor

— Height spurt

3. Stage three Penis enlarges

— Pubic hair starts growing

— Ejaculation (wet dreams) Breasts enlarge

— Pubic hair darkens, becomes curlier

— Vaginal discharge

4. Stage four Continued enlargement of testes and penis

— Penis and scrotal sac deepen in color

— Pubic hair curlier and coarser

— Height spurt

— Male breast development Onset of menstruation

— Nipple is distinct from areola

5. Stage five Fully mature adult

— Pubic hair extends to inner thighs

—Increases in height slow, then stop Fully mature female

— Pubic hair extends to inner thighs

— Increases in height slow, then stop

Genetic Factors

Perhaps more than any other factor, your youngster's pattern of growth and ultimate height will be influenced by heredity. Your son, for example, may want to be one of the tallest boys in his class, and he may aspire to play professional basketball. However, if both you and your spouse have below-average stature, his height as an adult will be more like yours than like his favorite sports idol's. While there are exceptions, tall parents usually have tall children, and short parents usually have short children. Those are the realities of genetics. Even so, if your child seems unusually short or tall relative to her friends of the same age, talk with your pediatrician. The doctor may recommend X-rays to determine your child's bone growth.

A true growth disorder can sometimes be treated by administering growth hormones; however, this therapy is reserved for youngsters whose own glands cannot produce this hormone, thus interfering with normal growth. Physicians do not recommend this treatment for healthy boys and girls who may want (or whose parents may want them) to grow to be 6 feet tall instead of 5 feet 8.

Boys' and Girls' Concerns About Puberty

Menstruation: Many concerns about puberty center on menstruation. Spend time helping your daughter prepare for her first period. There is no reason for a girl to be surprised by her menarche, not knowing what is happening or why. Remember, menstruation may begin sooner than you expect. Certainly, once your daughter's breast development has started, the two of you should fully discuss this topic. If you do not have adequate knowledge, ask your pediatrician to refer you to some informational sources. Some pediatricians schedule special educational visits at the time of puberty.

Discuss the biology of menstruation, describing it as a normal bodily process. Mention that her periods may be irregular, particularly in the beginning as her body adapts to rapid physiological changes. Also, let her know that several months before her first period, fluid may be secreted by glands within her vagina. This substance may be clear or white in color, and watery to thick in consistency. Tell her not to worry, and that this so-called physiologic leukorrhea is normal. Explain that she may experience some cramping before or during her periods. If the cramps become severe, her doctor may have some suggestions for alleviating them, perhaps with physical exercises or medication.

Of course, discuss hygiene related to menstrual cycles. Be certain your daughter has the supplies she will need for her first period. Since she may be away from home when that first period begins, discuss how to use pads or tampons. She should understand the need to change pads or tampons several times a day, and that tampons should not be worn overnight. Of course, girls can shower or bathe while menstruating. Many girls will ask if they can participate in activities such as swimming, horseback riding or physical education classes. Reassure your daughter that she can take part in normal activities while menstruating. Exercise can sometimes even ease the cramps associated with periods.

Breast Development: Some girls also have anxieties about breast development. For example, one breast usually begins to develop before the other. Explain to your daughter that as her breasts develop, it is quite normal for one to be somewhat larger than the other and that breast size is seldom symmetrical. Also, when a girl first notices the lump beneath one nipple, she may worry that this is cancer; reassure her that one breast is beginning to develop before the other, but if questions persist, consult your doctor. If a girl's breasts start to develop relatively early, she often feels embarrassed and self-conscious. To help your daughter feel more comfortable in a situation like this, she may prefer to wear loose-fitting clothing that disguises her early breast development. Also, be willing to buy her a "training bra" when she requests one.

Boys have pubertal concerns and worries, too, including:

Voice change: As their larynx (or voice box) enlarges and the muscles or vocal cords grow, their voice may "crack" as they speak. While this can be embarrassing and annoying, it's a normal part of the growth process.

Wet Dreams. Boys may awaken in the morning with damp pajamas and sheets. These "wet dreams," or nocturnal emissions, are caused by an ejaculation, not urination, that occurs during sleep; they are not an indication that the boy was having a sexual dream. Explain this phenomenon to your son, and reassure him that you understand that he cannot prevent it from happening. Wet dreams are just part of growing up.

Involuntary Erections: During puberty, boys get erections spontaneously, without touching their penis and without having sexual thoughts. These unexpected erections can be quite embarrassing, especially if they occur in public - at school, for example. Inform your son that these unexpected erections are normal and are a sign that his body is maturing. Explain that they happen to all boys during puberty, and that with the passage of time they will become less frequent.

Breast Enlargement: Many boys experience swelling of the breasts during the early years of puberty. Most often, your son may feel a flat, buttonlike bump under one or both nipples. His breasts may also feel tender or even painful. After a few months - sometimes longer - the swelling will disappear; these boys will not develop true breasts.

One Testicle Lower than the Other: Uneven testicles, although they may be embarrassing in the boys' locker room, are both normal and common.

Puberty Information For Boys

Girls are not the only ones who go through changes during puberty. Puberty for boys usually starts with a growth spurt at about 10 to 16 years of age. You may notice that you grow out of your clothes or shoes a lot faster than you used to. Don't worry, just as with girls, your hormones will balance out and your body will catch up.

Following are some other changes you will notice during puberty:

Body size: Arms, legs, hands, and feet may grow faster than the rest of your body. Until the rest of your body catches up, you may feel a little clumsy.

Body shape:You will get taller and your shoulders will get broader. You will gain a lot of weight. During this time, many boys experience swelling under their nipples. This may cause them to worry that they are growing breasts. If you experience this, don't worry. It is common among boys your age and is a temporary condition. If you are worried about it, talk to your pediatrician.

During puberty, your muscles will also get bigger. Try not to rush this part of your growth. You may have friends who work out with weights and equipment to build up muscles, and you may want to begin this type of training yourself - often before your body is ready for it. If you are interested in these activities, talk to your pediatrician about a safe time for you to begin weight training.

Voice: Your voice will get deeper. This may start with voice cracking. As you continue to grow, the cracking will stop and your voice will stay at the lower range.

Hair: Hair will appear under your arms, on your legs and face, and above your penis. Chest hair may appear during puberty or years after, although not all men have chest hair. Some men shave the hair on their faces. There is no medical reason to shave, it is simply a personal choice. If you decide to shave, be sure to use shaving cream and a clean razor made for men. It is a good idea to use your own personal razor or electric shaver and not to share one with your family or friends.

Skin: Skin may get more oily and you may notice you perspire more. This is because your glands are growing too. It is important to wash every day to keep your skin clean and to use a deodorant or antiperspirant to keep odor and wetness under control. Despite your best efforts to keep your face clean, you still may get pimples. This is called acne and is normal during this time when your hormone levels are high. Almost all teenage boys get acne at one time or another. Whether your case is mild or severe, there are things you can do to keep it under control. For more information on controlling acne, talk to your pediatrician.

Penis: Your penis and testes will get larger. You may have erections more often due to an increase in sex hormones. Erections occur when the penis gets stiff and hard - sometimes for no reason. This is normal. Even though you may feel embarrassed, try to remember that most people will not even notice your erection unless you draw attention to it. Many boys become concerned about their penis size; a boy may compare his own penis size with that of his friends. It is important to remember that the size of a man's penis has nothing to do with his manliness or sexual functioning.

Your body will also begin to produce sperm during puberty. This means that during an erection, you may also experience ejaculation. This occurs when semen (made up of sperm and other fluids) is released through the penis. This could happen while you are sleeping. You might wake up to find your sheets or pajamas are wet. This is called a nocturnal emission or "wet dream." This is normal and will stop as you get older.

Puberty Information for Girls

Puberty is the time in a girl's life when her body changes from that of a young girl to that of a woman. It is also the time when a girl becomes physically able to have babies. Although there is no "right" time for puberty to begin, it generally starts earlier for a girl than it does for a boy - usually between 9 and 13 years of age. This is why many girls are taller and may act more mature than boys for a few years until the boys catch up.

Following are some of the changes your body will go through during puberty:

Breasts: In most girls, puberty starts with breast growth. When your breasts start to develop, you may notice small, tender lumps under one or both nipples that will get bigger over the next few years. When breasts first begin to develop, it is not unusual for one breast to be larger than the other. However, as they develop, they will most likely even out before they reach their final size and shape.

As your breasts develop, you may need a bra. Some girls feel that wearing a bra for the first time is exciting - it is the first step toward becoming a woman! However, some girls feel embarrassed, especially if they are among the first of their friends to need a bra. If the people around you make a bigger deal of your first bra than you would like, try to remember that they do not mean to embarrass you, they are just proud of how much you have grown.

Hair: Soft hair will start to grow in the pubic area (the area between your legs). This hair will eventually become thick and very curly. You may also notice hair under your arms and on your legs. Many women shave this hair. There is no medical reason to shave, it is simply a personal choice. If you decide to shave, be sure to use a lot of soap and water and a clean razor made for women. It is a good idea to use your own personal razor or electric shaver and not to share one with your family or friends.

Body shape: Hips get wider and your waist will get smaller. Your body will also begin to build up fat in the stomach, buttocks, and legs. This is normal and gives your body the curvier shape of a woman.

Body size: Arms, legs, hands, and feet may grow faster than the rest of your body. Until the rest of your body catches up, you may feel a little clumsier than usual.

Skin: Skin may get more oily and you may notice you sweat more. This is because your glands are growing too. It is important to wash every day to keep your skin clean and to use a deodorant or antiperspirant to keep odor and wetness under control. Despite your best efforts to keep your face clean, you still may get pimples. This is called acne and is normal during this time when your hormone levels are high. Almost all teenagers get acne at one time or another. Whether your case is mild or severe, there are things you can do to keep it under control. For more information on controlling acne, talk to your pediatrician.

Menstruation: Your menstrual cycle, or "period," begins. Most girls get their periods between 9 and 16 years of age.

PIAGET THEORY OF DEVELOPMENT

Piaget was born in Switzerland in 1896, his studies of human cognitivedevelopment help us to understand how children percieve the world around them. Through his interest of children, he became one of this centuries mostinfluential observers of children. He worked for over fifty years observing andstudying children and their behaviors. Piaget realized that a child's mind wasnot a miniature version of the adult.

Originally, Piaget was trained as a biologist and graduated with adoctorate from Neuchatel University. Because of his mother's poor mentalhealth, Piaget became interested in psychopathology. In 1919, Piaget moved toParis and took courses in psychopathology, logic, epistemology, and thephilosophy of science. His studies led him to look for an area ofexperimentation, he was recommended to Dr. Theophile Simon, who with AlfredBinet developed the first intelligence test. Piaget became interested in whychildren repeatedly choose the same incorrect answers to the intelligence tests. This motivated him to begin his studies of cognitive development in children.

Piaget's research was extremely comprehensive, in fact, Piaget coined theterm "genetic epistemology" to describe his approach to the study ofknowledge. Piaget felt that psychology would provide the missing link betweenthe philosophical problem of epistemology and biological intelligence . Piagetis the author of at least fifty books, numerous articles, and has at least 30honorary doctorates.

Over the years, Piaget has taught courses in philosophy, psychology, andsociology at the universities of Geneva,

Neuchatel, Lausanne, and Sorbonne. In1955 Piaget organized his own international center for genetic epistemology inGeneva. While in Geneva, Piaget became a familiar snowy haired, pipe in mouth,bike riding, intellectual figure commonly seen around the streets anduniversity. Piaget retired from teaching in 1979, but continued his researchand work at the center until his death in 1980.

Intelligence

When considering intelligence, Piaget focuses on the mental processes thatoccur, rather than on the actual measure of the intellect. He uses four areasto define intelligence. These areas are: a biological approach to looking at intelligence, the succession of the stages, knowledge, and intellectualcompetence.

Piaget's biological approach, or biological adaption, focuses on thephysical and mental aspects of our bodies. This includes our reflexes whichoccur when certain stimuli trigger an instinctive response. He also discusseshow we adapt to certain situations using assimilation and accomodation. Assimilation occurs when new information is introduced to a person. The personbegins to integrate the new information into existing files, or "schema". Accomodation occurs when the person reorganizes schema to accomodatethemselves with the environment.

The succession of stages involves the movement through four stages thatPiaget has set and defined. Children must move through these stages duringtheir childhood. These include Sensorimotor, Preoperational, Concreteoperational, and Formal operational. Stage movement is an important factor ofPiaget's definition of intelligence, because Piaget states there are a specificset of criteria that must be met and mastered at each stage. In order to movefrom the first stage to the next, the child must master that specific set ofcriteria.

To define Intellectual Competence, Piaget focuses on the highest level offunctioning that can occur at a specific stage.

Although Piaget has approximateages assigned to stages, a child's competence is only measured by what stagethey are in, not by age. If the child can only perform, tasks that are at thepreoperational stage, that is the highest level the child is at regardless ofage.

Sensorimotor Stage

The sensorimotor stage is the first of the four stages Piaget uses to definecognitive development. Piaget designated the first two years of an infants lifeas the sensorimotor stage.

During this period, infants are busy discovering relationships betweentheir bodies and the environment. Researchers have discovered that infants haverelatively well developed sensory abilities. The child relies on seeing,touching, sucking, feeling, and using their senses to learn things aboutthemselves and the environment. Piaget calls this the sensorimotor stagebecause the early manifestations of intelligence appear from sensory perceptionsand motor activities.

Through countless informal experiments, infants develop the concept ofseperate selves, that is, the infant realizes that the external world is not anextension of themselves.

Infants realize that an object can be moved by a hand (concept ofcausality), and develop notions of displacement and events. An importantdiscovery during the latter part of the sensorimotor stage is the concept of "objectpermanence".

Object permanence is the awareness that an object continues to existeven when it is not in view. In young infants, when a toy is covered by a pieceof paper, the infant immediately stops and appears to lose interest in the toy(see figure above).This child has not yet mastered the concept of objectpermanence. In older infants, when a toy is covered the child will activelysearch for the object, realizing that the object continues to exist.

After a child has mastered the concept of object permanence, the emergenceof "directed groping" begins to take place. With directedgroping, the child begins to perform motor experiments in order to see what willhappen. During directed groping, a child will vary his movements to observe howthe results will differ. The child learns to use new means to achieve an end. The child discovers he can pull objects toward himself with the aid of a stickor string, or tilt objects to get them through the bars of his playpen.

The Preperational Stage

The preoperational stage occurs from 2 to 6 years of age, and is the secondstage in Piaget's stages of cognitive development. Throughout most of the preoperational stage, a child's thinking isself-centered, or *egocentric*. According to Piaget, during thepreoperational stage a child has difficulty understanding life from any otherperspective than his own.

Egocentrism is very apparent in the relationship between two preschoolchildren. Imagine two children are playing right next to each other, oneplaying with a coloring book and the other with a doll. They are talking toeach other in sequence, but each child is completely oblivious to what the otheris saying.

Julie: "I love my dolly, her name is Tina"

Carol: "I'm going to color the sun yellow"

Julie: "She has long, curly hair like my auntie"

Carol: "Maybe I'll color the trees yellow, too"

Julie: "I wonder what Tina's eyes are made of?"

Carol: "I lost my orange crayon"

Julie: " I know her eyes are made of glass."

These type of exchanges are called "collective monologues". Thistype of monolouge demonstrates the egocentrism of children's thinking in thisstage.

Egocentrism of the young child leads them to believe that everyone thinks asthey do, and that the whole world shares their feelings and desires. This senseof oneness with the world leads to the child's assumptions of magic omnipotence. Not only is the world created for them, they can control it. This leads to thechild believing that nature is alive, and controllable. This is a concept ofegocentrism known as *"animism"*, the most characteristic ofegocentric thought.

Closely related to animism is *artificialism*, or the idea thatnatural phenomena are created by human beings. Such as the sun is created by aman with a match. "Realism" is the child's notion that their ownperspective is objective and absolute. The child thinks from one perspectiveand regards this reality as absolute. Names, for example, are real to thechild. The child can't realize that names are only verbal labels, or conceivethe idea that they could have been given a different name.

In the pre-operational stage, the child begins to develop the use ofsymbols(but can not manipulate them), and the child is able to use language andwords to represent things not visible. Also, the pre-operational child beginsto master conservation problems.. Although the child is still unable to thinkin a truely logical fashion, the child may begin to treat objects as part of agroup.The pre-operational child may have difficulty with *classification*.

In the Latter part of the preoperational stage, the child begins to have anunderstanding between reality and fantasy. The child also begins to understandsex roles in society.

Concrete Operational Stage

The concrete operational stage is the third stage in Piaget's theory. Thisstage typically occurs between the ages of 7 and 12.

During this stage, the child begins to reason logically, and organizethoughts coherently. However, they can only think about acutal physicalobjects, they cannot handle abstract reasoning.

This stage is also characterized by a loss of egocentric thinking. During this stage, the child has the ability to master most types of conservationexperiments, and begins to understand reversibility. The concreteoperational stage is also characterized by the child's ability to coordinatetwo dimensions of an object simultaneously, arrange structures in sequence, andtranspose differences between items in a series

Formal Operational Stage

The formal operational stage is the fourth and final stage in Piaget'stheory. It begins at approximately 11 to 12 years of age, and continuesthroughout adulthood, although Piaget does point out that some people may neverreach this stage of cognitive development.

The formal operational stage is characterized by the ability to formulatehypotheses and systematically test them to arrive at an answer to a problem. The individual in the formal stage is also able to think abstractly and tounderstand the form or structure of a mathematical problem.

Another characteristic of the individual is their ability to reason contraryto fact. That is, if they are given a statement and asked to use it as thebasis of an argument they are capable of accomplishing the task. For example,they can deal with the statement "what would happen if snow were black"

Stage Movement

According to Piaget, there are four interrelated factors that allow movementfrom stage to stage. These factors include maturation, experience, socialinteraction, and equilibration. *Maturation* is the physical andpsychological growth that occurs in the child at a specific stage. *Experienceis* when the child thinks and interacts with real or concrete objects in theexternal environment. *Social interaction* involves the child socializingwith others, especially chilren. The last factor of stage movement is *equilibration*,this occurs when the child brings together maturation, experience, and socialinteraction in order to build mental schema. Equilibration is considered to bethe tendency for children to seek cognitive coherence and stability. They aremotivated in this drive for equilibration by *disequilibrium* or aperceived discrepancy between an existing scheme and something new.

Piagets Conservation Experiments

Conservation is the realization that quantity or amount does not changewhen nothing has been added or taken away from an object or a collection ofobjects, despite changes in form or spatial arrangement.

Jean Piaget used the idea of conservation in a set of experimentsgeared at studying children's ability to think and reason. Piaget realized thatmaturing children progressively master different types of conservation as theyreason about their world. The cow figure below represents conservation of area.When a child realizes that cows eat the same amount of grass regardless of thespatial location of the grass , they can conserve area.

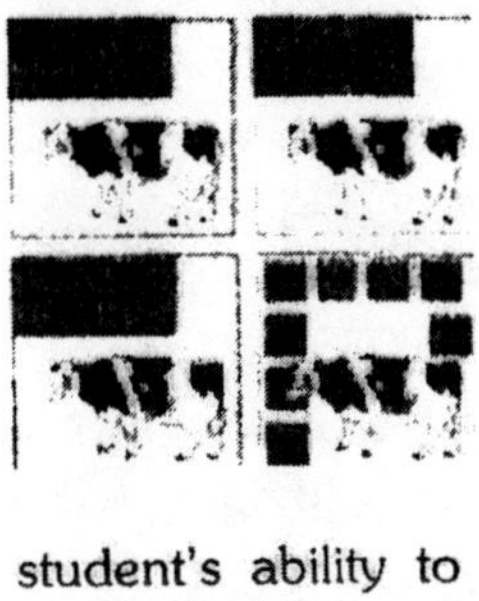

According to piaget, a student's ability to solve conservation problemsdepends on an understanding of three basic aspects of reasoning: identity,compensation, and reversability.

With mastery of *identity:* The student realizes that materialremains the same if nothing is added to or subtracted from the material.

With mastery of *compensation:* The student realizes that changesin one dimension can be offset by changes in another.

With mastery of *reversibility:* The student realizes that achange may be canceled out by mentally reversing the steps and returning to theorigin.

Piagetian Conservation Experiments

As suggested by Albert Einstein, Piaget studied a child's conception of time, space, and motion. Piaget published his results in two volumes: TheChild's Conception of Time and The Childs Conception of Movement and Speed.Itshould be recognized that Piaget went beyond a child's conception ofconservation and studied the child's conception of dimension and number.

Before a child can develop a true conception of numbers the useage ofone-to-one correspondence must be recognized by the child. Piaget's conservationof number experiment displayed two rows of counters placed in one-to-onecorrespondence . Subjects admit their equality. One of the rows is thenelongated or contracted. The subjects are asked whether each row has the samenumber of counters. Numbers can be reversed (by subtraction) or combined (byaddition) and express characteristics of combinativity, association, identity,and reversibility. The ability of a subject to solve this conservation problemdepends upon subjects mastery of identity. The average age that a child canconserve number is approximately 6-7 years of age.

Conservation of substance demonstrates the principle that alteration ofan object (malleable clay) doesn't change the quantity or mass of an object.Before a child can conserve substance, which is achieved at the age of 7-8, thechild experiences "perception dominance" in which the appearance ofsomething dominates over logical thought progression. Conservation of length isa classic example of "perception dominance", a length of rope is notchanged by an alteration in configuration of the rope.

Piaget used a geometrical experiment called "cows on a farm"to test for conservation of area. To illustrate this, Piaget used greencardboard to represent farmland. Two identical green farms were established,each had a little wooden cow placed upon it, subjects were asked whether thecows had the same amount of grass to feed upon. Piaget would proceed to equallyadd little cubic farmhouses to the models. In one farmyard the houses where allarranged in a tidy row, on the other farmland the houses were all scatteredabout. Until the ages of 8-9, children are irresistibly inclined to think thereis less grazing land on the farmland which contains the scattered houses.

Conservation of volume is usually tested using a cylinder and a flatdish (see figure below). Children are inclined to think that a tall, erect,narrow dish contains more liquid than an equal amount in a flat dish. Mastery ofconservation of volume requires reverse thinking and is the last of thePiagetian conservation tests children grasp.

A Biological Theory of Knowledge

Piaget never left biology, he had a lifelong interest in the plantsedum, varieties of which are common in many rock gardens. Similarily, Piagetstudied the snail Limnaea Stagnalis for years. Piaget transplanted sedum plantsinto the Savoy Alps in order to see how the plant reacted to coldertemperatures. The plants reacted by yeilding smaller, thicker leaves in orderto increase photosynthesis. He also transplanted the snails from tranquil watersto turbulent rocky shores, the snails reacted by changing their shells from anelongated shape to a globular shape.

Piaget felt that living organisms are self-regulating in their choicesof ways to adapt. This differs from Darwin who saw organisms as passive anddependent on mutation and millions of years of evolution for survival. Piagetfelt that adaptation was a continuous process in which the hereditary structureof the organism interacts with the environment in such a way as to reconstituteitself for better survival (Pulaski, 1980). Piaget saw a link between biologicalintelligence and human being intelligence.

Piagetian theory explains how human intelligence develops through anintellectual regulatory process geared by *adaptation* to the environment. Duringthis on-going relationship with the environment the child exhibits certainorganizations based upon *assimilation*- the taking in process of experience, accepting new encounters and fitting them into existing schemes, and *accommodation*- the reaction of the individual who encounters newexperiences that are not consistent with existing schemes and so the personmust change their scheme to accept or accommodate the new information.

Piaget felt that a baby is an active and curious organism, thatreaches out and seeks to regulate a balance between assimilation andaccomodation. This balance is what Piaget describes as equilibrium. Piaget considered the process of equilibrium an important factor in thecognitive growth and development of a child. It was for this reason Piagetinsisted that children must be allowed to do their own learning. Piaget realizedthat humans progressively develop or mature to higher states of cognitivedevelopment andrealized that children acquire knowledge transmitted by parents, teachers,and books, he called this "social transmission." Piaget believed thatwhen a child hears contradictory statements that challenge establishedschemes, equilibrium is disturbed.

Piaget called such a disruption inequilibrium "cognitive conflict or disequilibrium." When children experience cognitiveconflict they set out in search of an answer that will enable them to acheivestates of equilibrium. Piaget felt that all children go through certainstages of intellectual development in the same order, even though thechronological ages may vary between bright and dull students.

6

Psychotherapy and Child Development

Adolescence is a time of major change in a person's life. During this period there are several drastic changes in the areas of physical, intellectual, emotional, and social growth. In order to fully understand this group of individuals, the changes that they are experiencing need to be understood first.

Physical Development

— Both boys and girls experience a growth spurt

— Sexual organs rapidly develop

— Sweat glands become more active and acne may appear

— Feel tired more easily

— Require a greater amount of calories

These drastic changes in the physical make up of the adolescents' body may lead to an awkwardness and self-consciousness. Adolescents can begin to see themselves very differently than they ever have before. This may cause them to begin comparing themselves with their peers. It can be very frustrating for the adolescents who see themselves as not developing as rapidly as their peers seem to be. This puts them at a greater risk for fad medicines and diets.

Intellectual Development

- 85% of 12-14 year olds experience a plateau in brain development
- Most all 10-14 year olds make progress in intellectual development, but the gap between good and poor students widens
- Feel need to be competent and gain further achievement
- Begin to develop formal operational reasoning
- Show growth in ability to reason and generalize after observing and analyzing data
- Improve ability to function creatively and effectively
- Gain greater intellectual curiosity
- Attention span grows
- Feel importance of planning for rest of life
- Between 10-15 years old largely decide attitudes about learning and work as well as adult values

Adolescence can be a very frustrating time in life for many people. They can begin to feel as if they are not as smart as their peers. This frustration is enlarged if the demands of intellectual growth in school continue to increase while the student feels like he/she is getting left behind. When their need to feel competent and gain further achievement is not attained the person can show frustration or retreat to helplessness. It is during this stage in life that people begin to assume a greater responsibility for self-direction.

Adolescents also make a large amount of decisions about how to make and spend money. These two changes can be the root of several home life problems if the adolescent's parents do not agree with the decisions that the adolescent makes. Adolescents begin to like to see concrete outcomes from their efforts and also enjoy the opportunity for creative expression. This group of individuals needs to learn decision-making through experience rather than someone telling them which way to do things to achieve the best outcome. This concept is often a challenging one for adults to realize and accept.

Emotional Development

- — Move away from parental influence toward greater independence
- — Begin to criticize opinions, suggestions, and actions of adults
- — Frequently fluctuate between emotional peaks of excitement and depths of moodiness
- — Isolation from and lack of involvement in the community leads to feelings of being ignored and not needed
- — Can be very unstable
- — Often make rude comments that can hinder another's self-concept
- — Learn to feel, think, and act according to congruent sex roles
- — Develop further interest in ethics and morality and form own opinions about these issues

Dr. Karl Menninger, an early American psychiatrist, has observed that today's children are not getting their belonging needs fulfilled at home, at school, or from neighborhoods and so are therefore striving to obtain "artificial belongings". Belonging will be found only in relationships with adults who see, as Menninger put it, that "living with and loving other human beings who return that love is the most strengthening and salubrious emotional experience in the world". If they do not find these places of belonging within the school or at home they will find them outside these facilities such as at the local mini mart.

Even thought it is very important that adolescents have a sense of belonging somewhere it is also crucial that they have a sense of independence. Without this they may begin to see themselves as just a pawn in a world where others control their destiny. This can lead to many self-destructive behaviors such as learned helplessness, absence of an internal locus of control, and lack of intrinsic motivation. They begin to criticize opinions, suggestions, and actions of adults, probably to establish their own ideas.

Social Development

— Seek out more privacy

— Conversation with family is less spontaneous

— Peers begin to move into their life and take the place of family

— Same sex relationships tend to get stronger as they try to "fit in" to a crowd

— Girls tend to have more intimate and exclusive friendships than boys

— Seek out places to congregate with their interest mates

It is during this period of life that the need to be accepted by peers becomes the most important. It is crucial to the social development of adolescents that they feel like they fit in and are needed by someone. This does not mean that adults can not play a role in the individual's life at this stage of development. It just emphasizes the fact the adults need to recognize these needs and help guide them to acceptable places where they can gain this social need.

ADOLESCENT GROWTH AND DEVELOPMENT

Adolescence is a time of many transitions both for teens and their families. To ensure that teens and adults navigate these transitions successfully, it is important for both to understand what is happening to the teen physically, cognitively, and socially; how these transitions affect teens; what adults can do; and what support resources are available. As you read the following information, keep in mind that while all teens develop, they don't all follow the same timeline.

Physical Development

During the teen years, adolescents experience changes in their physical development at a rate of speed unparalleled since infancy. Physical development includes:

- *Rapid gains in height and weight*: During a one-year growth spurt, boys and girls can gain an average of 4.1 inches and 3.5 inches in height respectively. This spurt typically occurs two years earlier for girls than for boys. Weight gain results from increased muscle development in boys and body fat in girls.
- *Development of secondary sex characteristics*: During puberty, changing hormonal levels play a role in activating the development of secondary sex characteristics. These include:
  (1) growth of pubic hair;
  (2) menarche (first menstrual period for girls) or penis growth (for boys);
  (3) voice changes (for boys);
  (4) growth of underarm hair;
  (5) facial hair growth (for boys); and
  (6) increased production of oil, increased sweat gland activity, and the beginning of acne.
- *Continued brain development*: Recent research suggests that teens' brains are not completely developed until late in adolescence. Specifically, studies suggest that the connections between neurons affecting emotional, physical and mental abilities are incomplete. This could explain why

some teens seem to be inconsistent in controlling their emotions, impulses, and judgments.

How do these changes affect teens?

— Teens frequently sleep longer. Research suggests that teens actually need more sleep to allow their bodies to conduct the internal work required for such rapid growth. On average, teens need about 9 1/2 hours of sleep a night.

— Teens may be more clumsy because of growth spurts. If it seems to you that teens' bodies are all arms and legs then your perception is correct. During this phase of development, body parts don't all grow at the same rate. This can lead to clumsiness as the teen tries to cope with limbs that seem to have grown overnight. Teens can appear gangly and uncoordinated.

— Teenage girls may become overly sensitive about their weight. This concern arises because of the rapid weight gain associated with puberty. Sixty percent of adolescent girls report that they are trying to lose weight. A small percentage of adolescent girls (1-3%) become so obsessed with their weight that they develop severe eating disorders such as anorexia nervosa or bulimia. Anorexia nervosa refers to starvation; bulimia refers to binge eating and vomiting.

— Teens may be concerned because they are not physically developing at the same rate as their peers. Teens may be more developed than their peers ("early-maturers") or less developed than their peers ("late-maturers"). Being out of developmental "step" with peers is a concern to adolescents because most just want to fit in. Early maturation affects boys and girls differently. Research suggests that early maturing boys tend to be more popular with peers and hold more leadership positions. Adults often assume that early maturing boys are cognitively mature as well. This assumption can lead to false expectations about a young person's ability to take on increased responsibility. Because of their physical appearance, early maturing girls are more likely to experience pressure to become involved in dating relationships with older boys before they are emotionally ready. Early maturing girls tend to suffer more from depression, eating disorders, and anxiety.

— Teens may feel awkward about demonstrating affection to the opposite sex parent. As they develop physically, teens are beginning to rethink their interactions with the opposite sex. An adolescent girl who used to hug and kiss her dad when he returned home from work may now shy away. A boy who used to kiss his mother good night may now wave to her on his way up the stairs.

— Teens may ask more direct questions about sex. At this stage, adolescents are trying to figure out their sexual values. Teens often equate intimacy with sex. Rather than exploring a deep emotional attachment first, teens tend to assume that if they engage in the physical act, the emotional attachment will follow. They may ask questions about how to abstain without becoming embarrassed or about how they will know when the time is right. They may also have specific questions about methods of birth control and protection from sexually transmitted diseases.

Knowledge about what changes and behaviors during adolescence are normal can go a long way in helping both teens and adults manage the transition successfully. There are also some specific things adults can do to be supportive:

— *Don't criticize or compare the teens to others.* Teens are already acutely self-conscious about the way they look. They don't need you to point it out to them.

— *Encourage teens to get enough sleep.* Realize they may need an extra boost in getting out of bed for school. Try to be understanding when teens want to sleep until noon on Saturday.

— *Encourage and model healthy eating habits*: Keep plenty of nutritious foods in the house. Remember that teens need to take in more calories to fuel their growth. Monitor eating habits accordingly.

— *Encourage and model physical activity*: Exercise will help teens burn excess energy, strengthen developing muscles, and sleep better at night. It may also help teens become more comfortable in their changing bodies.

— *Provide honest answers to teens about sex*: Teens are in search of knowledge on this subject. If adults do not provide accurate information, teens are forced to rely on their peers or other potentially inaccurate sources. Unfortunately, such erroneous information is often to blame when teens make poor decisions.

— *Be understanding of their need for physical space*: Do not take it personally if your teen is not as physically affectionate as he or she was in the past. Do not force your teen to hug or kiss relatives or family friends. Maintain communication, but respect teens' need to withdraw.

— *Be patient with excessive grooming habits*: Teens often spend large amounts of time grooming themselves and obsessing over skin care products. Often, this behavior merely reflects teens' attempts to maintain some sense of control over their rapidly changing bodies.

Cognitive Development

Most adults recognize that teens have better thinking skills than younger youth. These advances in thinking can be divided into several areas:

— *Developing advanced reasoning skills*: Advanced reasoning skills include the ability to think about multiple options and possibilities. It includes a more logical thought process and the ability to think about things hypothetically. It involves asking and answering the question, "what if...?".

— *Developing abstract thinking skills*: Abstract thinking means thinking about things that cannot be seen, heard, or touched. Examples include things like faith, trust, beliefs and spirituality.

— Developing the ability to think about thinking in a process known as "meta-cognition." Meta-cognition allows individuals to think about how they feel and what they are thinking. It involves being able to think about how one is perceived by others. It can also be used to develop strategies, also known as mnemonic devices, for improving learning. Remembering the notes on the lines of a music staff (e, g, b, d, and f) through the phrase "every good boy does fine" is an example of such a mnemonic device.

How do these changes affect teens?

— Teens demonstrate a heightened level of self-consciousness. Teens tend to believe that everyone is as concerned with their thoughts and behaviors as they are. This leads teens to believe that they have an "imaginary audience" of people who are always watching them.

— Teens tend to believe that no one else has ever experienced similar feelings and emotions. They may become overly dramatic in describing things that are upsetting to them. They may say things like "You'll never understand," or "My life is ruined!"

— Teens tend to exhibit the "it can't happen to me" syndrome also known as a "personal fable." This belief causes teens to take unnecessary risks like drinking and driving ("I won't crash this car"), having unprotected sex (I can't possibly get pregnant), or smoking (I can't possibly get cancer").

— Teens tend to become very cause-oriented. Their activism is related to the ability to think about abstract concepts. After reading about cruelty to animals a teen may become a vegetarian and a member of "People for the Ethical Treatment of Animals" (P.E.T.A.). Another teen may become active in "Green Peace" or "Save the Whales" campaigns.

— Teens tend to exhibit a "justice" orientation. They are quick to point out inconsistencies between adults' words and their actions. They have difficulty seeing shades of gray. They see little room for error.

What can you do?

— Don't take it personally when teens discount your experience. Try to empathize with and listen to their concerns. Enlist the help of a slightly older sibling or friend to give good advice to the teen if needed.

— Get teens involved in discussing their behavioral rules and consequences. Teens should take a more active role in determining how they should behave. Their advanced reasoning skills make it easier for them to generate realistic consequences for their actions. Listen to their ideas!

— Provide opportunities for teens to participate in controlled risky behavior. Get teens involved in properly supervised extreme sports, such as parachuting, or rock climbing. Such activities will allow teens opportunities to play out their "it can't happen to me" mentality in an environment that won't be deadly if they fail.

— Provide opportunities for teens to get involved in community service. Teens want to become active in things that have deeper meaning. Suggest they volunteer at a homeless shelter, walk dogs for the animal shelter, or take meals to the elderly. Talk with them about their experiences.

— Talk to teens about their views and be open to discussing your own. Find out what they think about news stories on television or in the paper; ask them about their political and spiritual beliefs. Teens are already thinking about these things so give them a non-threatening forum for discussing them.

— Try to build a genuine relationship with your teen. Let them know what you were like as a teen. Talk to them about your mistakes and vulnerabilities. Try to understand their feelings and express yours so you can be understood.

Psycho-Social Development

There are five recognized psychosocial issues that teens deal with during their adolescent years. These include:

— *Establishing an identity*: This has been called one of the most important tasks of adolescents. The question of "who am I" is not one that teens think about at a conscious level. Instead, over the course of the adolescent years, teens begin to integrate the opinions of influential others (e.g. parents, other caring adults, friends, etc.) into their own likes and dislikes. The eventual outcome is people who have a clear sense of their values and beliefs, occupational goals, and relationship expectations. People with secure identities know where they fit (or where they don't want to fit) in their world.

— *Establishing autonomy*: Some people assume that autonomy refers to becoming completely independent from others. They equate it with teen "rebellion." Rather than severing relationships, however, establishing autonomy during the teen years really means becoming an independent and self-governing person within relationships. Autonomous teens have gained the ability to make and follow through with their own decisions, live by their own set of principles of right and wrong, and have become less emotionally dependent on parents. Autonomy is a necessary achievement if the teen is to become self-sufficient in society.

— *Establishing intimacy*: Many people, including teens, equate intimacy with sex. In fact, intimacy and sex are not the same. Intimacy is usually first learned within the context of same-sex friendships, then utilized in romantic relationships. Intimacy refers to close relationships in which people are open, honest, caring and trusting. Friendships provide the first setting in which young people can practice their social skills with those who are their equals. It is with friends that teens learn how to begin, maintain, and terminate relationships, practice social skills, and become intimate.

— *Becoming comfortable with one's sexuality*: The teen years mark the first time that young people are both physically mature enough to reproduce and cognitively advanced enough to think about it. Given this, the teen years are the prime time for the development of sexuality. How teens are educated about and exposed to sexuality will largely determine whether or not they develop a healthy sexual identity. More than half of most high school students report being sexually active. Many experts agree that the mixed messages teens receive about sexuality contribute to problems such as teen pregnancy and sexually transmitted diseases.

— *Achievement*: Our society tends to foster and value attitudes of competition and success. Because of cognitive advances, the teen years are a time when young people can begin to see the relationship between their current abilities and plans and their future vocational aspirations. They need to figure out what their achievement preferences are-what they are currently good at and areas in which they are willing to strive for success.

How do these changes affect teens?

— Teens begin to spend more time with their friends than their families. It is within friendship groups that teens can develop and practice social skills. Teens are quick to point out to each other which behaviors are acceptable and which are not. It is important to remember that even though teens are spending increased amounts of time with their friends, they still tend to conform to parental ideals when it comes to decisions about values, education, and long-term plans.

— Teens may have more questions about sexuality. They may ask about adults' values and beliefs. They may ask how you knew it was time to have sex or why you waited.

— Teens may begin to keep a journal. Part of achieving identity is thinking about one's thoughts and feelings. Teens often begin journaling as a way of working through how they feel.

— When they are in their rooms, teens may begin to lock their bedroom doors. Locking doors is a way to establish privacy. As long as teens continue to interact with the family, locked doors are usually nothing to worry about.

— Teens may become involved in multiple hobbies or clubs. In an attempt to find out what they are good at, teens may try many activities. Teens' interests also change quickly. Today they are into yoga, and tomorrow they are into soccer.

— Teens may become elusive about where they are going or with whom. When asked what they'll be doing for the evening, teens typically reply with "nothing" or "hanging out." When asked whom they'll be with, teens reply, "just some friends."

— Teens may become more argumentative. Teens may question adults' values and judgments. When teens don't get their way, they may say, "you just don't understand."

— Teens may not want to be seen with parents in public. They may make parents drop them off a block from their friends' houses or from school.

— Teens may begin to interact with parents as people. Even though they may not want to be seen with parents in public, teens may begin to view parents more as people. They may ask more questions about how a parent was when he or she was a teen. They may attempt to interact with adults more as equals.

What can you do?

— Encourage involvement in multiple groups or activities both within school and after-school. Realize that teens are trying to gain a sense of achievement-a sense of being uniquely good at something. Don't get frustrated if they frequently change their minds. At the same time, encourage them to stick with a project or activity long enough to establish some skills.

— Praise teens for their efforts as well as their abilities. This will help teens to stick with activities instead of giving up if they are not immediately successful.

— Help teens explore career goals and options. Take teens to work so they can see what adults do. Set up opportunities for them to "job shadow" others. Ask them questions about their future career goals. Remember that figuring out what they don't want to do is just as important as figuring out what they like!

— Give teens an opportunity to establish their behavioral guidelines and consequences. Allow teens to have input into curfew and other family rules. Their advanced cognitive skills coupled with their need for autonomy makes this a perfect time for them to provide suggestions and to demonstrate responsibility for their own behavior.

— Establish rituals to mark significant passages. Few rituals in our modern society mark the passage of teens to adulthood. Have a mother-daughter luncheon when the daughter gets her first period. Have a father-son outing when the son begins to shave. Have a family celebration when the teen moves from junior high to high school. Celebrate the teen's first driver's license and his or her ability to vote.

— Be aware of who your teens' friends are and what they are doing. Such parental monitoring should not end when youth enter their teen years. Despite teens' objections, make sure you know who their friends are and where they are going. Meet the parents of teens' friends. Provide fun things to do at home to encourage teens to "hang out" at your house so you'll know where they are and what they are doing.

— Continue to provide a structured environment. Teens should be allowed to have more independence, but not enough to place them in jeopardy. Despite their complaints, teens rely on adults to provide them with the sense of safety and structure they need to deal effectively with all the psychosocial tasks of adolescents.

MORAL DEVELOPMENT AND CRIME

Moral development, or "age" and "stage" theories are found in a number of fields, in general psychology, in cognitive psychology, and in adolescent psychology. This lecture will touch on the moral development ideas of Piaget, Erikson, Kohlberg, and Gilligan in the context of theories about normal and abnormal adolescence. It should be noted that despite the inroads of certain ideas, moral development theories are regarded as "eclectic" theories by criminology. Eclectic means "everything causes delinquency or crime." They're subject to what is called the "evil causes evil fallacy." Blaming delinquency on being an adolescent is the same as saying "moods" or "states of mind" cause crime (Neo-cognitive theory).

Adolescence as a Stage of Development

It's been called the "transition" stage with no accomplishment nor completion. Its existence has been called an "historical accident." Experts can't even agree on when it begins (most simply use the age of puberty) and when it ends (some say an "extended" adolescence is possible up to age 25 or 30; other say there's an "early" 12-14 stage, a "middle" 15-17 stage, and a "late" 18-19 stage). There's more mythology than science when it comes to notions of "typical" or "normal" adolescence, and explaining the breakdown that occurs.

Daedalus, an inventor, and his son, Icarus (so the ancient Greek myth goes) were both imprisoned in a tower on the island of Crete. To escape, Daedalus built a set of wings out of feathers and wax, but only his son, Icarus was small enough to fit thru the tower window. Daedalus warned Icarus not to fly too high because Apollo, God of the Sun, would become jealous of someone entering his domain and melt the wax. But Icarus, exhilarated by being able to soar above the Earth, flies too high, and his wings melt, plunging him into the Hellspont. Ever since, "adolescents" have been seen as:

(a) disobedient toward parents;

(b) wanting to "fly" too high; and, in some interpretations,

(c) condemned to hell.

Hippocrates, the father of medicine, recommended two kinds of treatments for unruly children. For females, they should be made pregnant. For males, they should be flogged. Philo, a friend of Socrates, advocated infanticide if beatings didn't work. History actually records a long record of using sex, corporal punishment, and infanticide against unruly children. The common law doctrine of in loco parentis has deep roots in almost all civilizations. Historically, adolescence can be considered a by-product of the Industrial Revolution. Prior to child labor laws, there had been no need to define any special period of life for those too young to work but too old to not be given at least some adult responsibilities.

Over the years, the age at which society permits children to assume adult roles has gradually been raised. This "stretch" of time, then, has come to be associated with the "stress" of adolescence. By the late 19th Century, numerous experts were popularizing the idea of adolescence being a "critical" stage of life. Baden-Powell (founder of the Boy Scouts) pioneered the technique of "not too much" but "not too little" discipline as a cure for the problems of adolescence. Dr. Spock, later in the 1950s, would advocate complete and total leniency. Counterculture experts such as Keniston and Roscak in the 1960s also advocated toleration as well as popularized the idea of being "stuck in adolescence." Freudian and Neo-Freudian ideas abounded, but it was Stanley Hall's ideas about the Six Themes of Adolescence that became the most popular hallmarks of "normal".

Themes of Adolescence

1. *Self-definition*: a concern for finding one's "real" self, a concern for habit formation, and little unstructured time to deal with these concerns, hence, a constant state of urgency.

2. *Estrangement and Omnipotentiality*: estrangement takes the form of feeling like a marginal member of society, with unrealistic mobility aspirations and intense job anxieties. Omnipotentiality is the feeling of absolute freedom, a reveling in all the pure possibilities that the future may hold. For many, the "car" becomes a symbol for this.

3. *Refusal of Socialization*: a critical if not rebellious stance at continued efforts to instruct, educate, or train for society's purposes. There's a feeling of always being under observation by a critical audience of all adult socializing agents.

4. *Celebration of Youth Culture*: a rebellious sense of solidarity based on the perceived sharing of fads, fashions, and styles by others in the same age group or generation; intense age-consciousness.

5. *Stasis as Death*: stasis means "standing still" or being in a rut, and this is avoided and despised at all costs. There's an irrational devotion to change, to putting oneself through changes merely for the sake of change. It takes two forms: a need to move (geographic restlessness); and a need to be moved (experimentation with states of consciousness).

6. *Physical Obsession*: there's an obsessiveness or inadmissible sense of shame over uncontrollable physical changes, like sex fantasies, body weight and contours, dietary habits, the outgrowing of clothing, outbreaks of acne, etc.

Indicators of normal and Abnoramal Adolescence

Indicators of being on a "normal" course of moral development:

1. Obsessive concern for bodily appearance

2. Fear of abandonment expressed as assertion of independence
3. Desire to be different in terms of "fads"
4. Sexual desire and manipulativeness
5. Wanting to be like other races or cultures
6. Persistent wisecracking as long as it's witty
7. Obsessive desire for success and recognition
8. Lack of self-identity or distinct self-concept
9. Emotional extremes expressed as sensitivity to criticism

Indicators of being on an "abnormal" course of moral development:

1. Driven by whim or caprice rather than purpose or gain
2. Unmoved by overtures of help & harms helpers
3. Shows no loyalty to other adolescents
4. Words are inconsistent with feelings, language is strange, humor missing
5. Claims to have always been the first to do something
6. Pathological lying for no good reason
7. Superficially charming but unable to maintain intimate relations

Indicators of being an "at-risk" youth (Middle Schools):

1. Frequent absenteeism, tardiness, or suspension
2. Academic performance below grade level or repeating a grade
3. Oppositional stance towards authority
4. Drug/alcohol involvement
5. Police/probation involvement

6. Being a "latch-key" child
7. Coming from a single parent family or foster home

Indicators of "delinquent pathways" (Justice Dept.):

1. *Authority Conflict Pathway*: Early signs of stubbornness, outright defiance by age 11, running away, avoiding authorities, truancy, staying out late
2. *Covert Pathway*: Early signs of minor mischievousness, frequent lying, shoplifting by age 10, property damage, firestarting, vandalism
3. *Overt Pathway*: Early acts of aggression or bullying, annoying everyone around them by age 12, having no friends and numerous enemies, interested only in escalating problems

Levels of moral development

Level 1: *preconventional morality*

Stage 1: Individual obeys rules in order to avoid punishment.

Stage 2: Individual conforms to society's rules in order to receive rewards.

Level 2: *conventional morality*

Stage 3: Individual behaves morally in order to gain approval from other people.

Stage 4: Conformity to authority to avoid censure and guilt.

Level 3: *postconventional morality*

Stage 5: Individual is concerned with individual rights and democratically decided laws.

Stage 6: Individual is entirely guided by his or her own conscience.

7

Role of Society in Overall Development of Children

ROLE OF PARENTS

As a child leaves infancy and approaches toddlerhood, one of the tasks parents face is introducing the child to the peer group. To be sure, parents are interested in their child's earliest interactions with peers, but in time, parents become more seriously invested in their children's ability to get along with playmates. *Getting along* has different meanings for different parents, but in general, parents want their child to enjoy the company of other children, be liked by them, be well-behaved in their presence (for example, share and cooperate with them), and resist the influence of companions who are overly boisterous, aggressive or defiant of adult authority.

How do parents help their child become a socially competent, well-liked playmate who is not too easily influenced by ill-behaved peers? What do we know from research literature in this area? Inasmuch as peer relations is only one of many social relationships that a child must master, it is not surprising that research on parenting styles gives some helpful insights into development of social skills in the peer group. A number of investigators, such as Diana Baumrind, Martin Hoffman, and Martha Putallaz, have made significant contributions to this topic.

The research of Diana Baumrind is particularly noteworthy. Baumrind has published a series of studies on the relation between parental child rearing styles and social competence in children of preschool and school age. Data on nursery school children were obtained from observations in a school setting and in laboratory test situations when the children were approximately four to five years of age. Data on the children's parents were obtained through home observations and interviews of both mothers and fathers. Three contrasting parenting styles were identified by Baumrind: authoritarian, permissive, and authoritative, each of which has implications for the child's social competence with peers and adults.

The three parenting styles differ particularly on two parenting dimensions: the amount of nurturance in child-rearing interactions and the amount of parental control over the child's activities and behavior. Authoritarian parents tend to be low in nurturance and high in parental control compared with other parents. They set absolute standards of behavior for their children that are not to be questioned or negotiated. They favor forceful discipline and demand prompt obedience. Authoritarian parents also are less likely than others to use more gentle methods of persuasion, such as affection, praise and rewards, with their children.

Consequently, authoritarian parents are prone to model the more aggressive modes of conflict resolution and are lax in modeling affectionate, nurturant behaviors in their interactions with their children. In sharp contrast, permissive parents tend to be moderate to high in nurturance, but low in parental control. These parents place relatively few demands on their children and are likely to be inconsistent disciplinarians. They are accepting of the child's impulses, desires, and actions and are less likely than other parents to monitor their children's behavior. Although their children tend to be friendly, sociable youngsters, compared with others their age they lack a knowledge of appropriate behaviors for ordinary social situations and take too little responsibility for their own misbehavior.

Authoritative parents, in contrast to both authoritarian and permissive parents, tend to be high in nurturance and moderate in parental control when it comes to dealing with child behavior. It is this combination of parenting strategies that Baumrind and others find the most facilitative in the development of social competence during early childhood and beyond. The following discussion describes specific behaviors used by authoritative parents and the role these behaviors play in fostering social development.

High Nurturance

Nurturing behaviors of parents that predict social competence include affectionate and friendly interaction with the child; consideration for the child's feelings, desires and needs; interest in the child's daily activities; respect for the child's points of view; expression of parental pride in the child's accomplishments; and support and encouragement during times of stress in the child's life. The advantages of high levels of nurturance in fostering social development have been confirmed again and again in studies of children. These advantages begin in infancy, when maternal nurturance facilitates a secure attachment which, in turn, predicts social competence, and continue throughout childhood.

High levels of nurturance in child rearing virtually assure more positive adult-child interactions than negative ones in the day-to-day operations of family life. This, in turn, predisposes the child to return love to the parent and to enjoy spending time with the parent, thus increasing the possibilities of significant parental influence throughout childhood. Parental nurturance also motivates the child to please the parent by striving to live up to parental expectations and helps to keep the child from hurting or disappointing the loved parent. Because children more readily identify with nurturant than nonnurturant models, the children of nurturing parents are more likely to incorporate parental values, such as considerateness and fairness in interpersonal relations, into their own lifestyle.

One would also expect these children to resist peer group values that are clearly different from family values. If there is a downside to high levels of nurturance in child rearing, it is the risk that nurturant parents might be more lax than other parents in challenging their children to measure up to developmentally appropriate standards for behavior. This risk would appear to be reduced, however, by the authoritative parents' inclination to combine moderate levels of parental control with nurturance.

Moderate Control

Nurturant parents who maintain at least a moderate level of control over their child do not give up their right to set behavioral standards for the child and to convey the importance of compliance with those standards. To facilitate compliance, and as a courtesy to the child, authoritative parents offer reasons and explanations for the demands placed on their children. Evidence suggests that such a practice increases the child's understanding of rules and regulations, eventually making it possible for the child to monitor his or her behavior in the absence of the parent. Parents who use authoritative child rearing practices often use positive reinforcers, such as praise, approval, and rewards, to increase the child's compliance with behavioral standards.

The success of positive social reinforcement in producing desirable behavior is legendary. A parent's positive response to good behavior may be the most powerful tool the parent has for increasing child compliance and decreasing the need for disciplinary action. When misbehavior does occur and discipline is deemed necessary, authoritative parents show a preference for "rational-inductive discipline," in which both sides of an issue are stated and a just solution is sought. These parents also prefer "consequence-oriented discipline" in which children are expected to make up for their wrongdoing.

Martin Hoffman points out that this disciplinary strategy has the advantage of focusing the child's attention on the plight of the victim rather than on the child's plight at the hands of an angry parent. Finally, authoritative parents try to avoid the more extreme forms of punishment in rearing their children. They do not favor harsh physical punishment or put-downs, such as ridicule or negative social comparison, which attack the child's sense of personal worth. Although the harsher forms of punishment can be effective in the short run, they often generate resentment and hostility that carry over to the school and peer group, reducing the child's effectiveness in these settings.

In parenting, as in other endeavors, nothing works all of the time. It is safe to say, however, that authoritative parenting works better than most other parenting styles in facilitating the development of social competence in children at home and in the peer group. High levels of nurturance combined with moderate levels of control help adults be responsible child rearing agents for their children and help children become mature, competent members of society. With a little bit of luck, the children of authoritative parents should enjoy more than their share of success in the peer group.

ROLE OF TEACHERS—PRESCHOOL, KINDERGARTEN, ELEMENTARY, MIDDLE, AND SECONDARY

Teachers act as facilitators or coaches, using interactive discussions and "hands-on" approaches to help students learn and apply concepts in subjects such as science, mathematics, or English. They utilize "props" or "manipulatives" to help children understand abstract concepts, solve problems, and develop critical thought processes. For example, they teach the concepts of numbers or of addition and subtraction by playing board games. As the children get older, the teachers use more sophisticated materials, such as science apparatus, cameras, or computers. To encourage collaboration in solving problems, students are increasingly working in groups to discuss and solve problems together.

Preparing students for the future workforce is the major stimulus generating the changes in education. To be prepared, students must be able to interact with others, adapt to new technology, and think through problems logically. Teachers provide the tools and the environment for their students to develop these skills. Preschool, kindergarten, and elementary school teachers play a vital role in the development of children. What children learn and experience during their early years can shape their views of themselves and the world and can affect their later success or failure in school, work, and their personal lives.

Preschool, kindergarten, and elementary school teachers introduce children to mathematics, language, science, and social studies. They use games, music, artwork, films, books, computers, and other tools to teach basic skills. Preschool children learn mainly through play and interactive activities. Preschool teachers capitalize on children's play to further language and vocabulary development (using storytelling, rhyming games, and acting games), improve social skills (having the children work together to build a neighborhood in a sandbox), and introduce scientific and mathematical concepts (showing the children how to balance and count blocks when building a bridge or how to mix colors when painting).

Thus, a less structured approach, including small-group lessons, one-on-one instruction, and learning through creative activities such as art, dance, and music, is adopted to teach preschool children. Play and hands-on teaching also are used in kindergarten classrooms, but there academics begin to take priority. Letter recognition, phonics, numbers, and awareness of nature and science, introduced at the preschool level, are taught primarily by kindergarten teachers. Most elementary school teachers instruct one class of children in several subjects. In some schools, two or more teachers work as a team and are jointly responsible for a group of students in at least one subject.

In other schools, a teacher may teach one special subject—usually music, art, reading, science, arithmetic, or physical education—to a number of classes. A small but growing number of teachers instruct multilevel classrooms, with students at several different learning levels. Middle school teachers and secondary school teachers help students delve more deeply into subjects introduced in elementary school and expose them to more information about the world. Middle and secondary school teachers specialize in a specific subject, such as English, Spanish, mathematics, history, or biology. They also can teach subjects that are career oriented.

Vocational education teachers, also referred to as career and technical or career-technology teachers, instruct and train students to work in a wide variety of fields, such as healthcare, business, auto repair, communications, and, increasingly, technology. They often teach courses that are in high demand by area employers, who may provide input into the curriculum and offer internships to students. Many vocational teachers play an active role in building and overseeing these partnerships. Additional responsibilities of middle and secondary school teachers may include career guidance and job placement, as well as followups with students after graduation.

Teachers may use films, slides, overhead projectors, and the latest technology in teaching, including computers, telecommunication systems, and video discs. The use of computer resources, such as educational software and the Internet, exposes students to a vast range of experiences and promotes interactive learning. Through the Internet, students can communicate with students in other countries. Students also use the Internet for individual research projects and to gather information. Computers are used in other classroom activities as well, from solving math problems to learning English as a second language. Teachers also may use computers to record grades and perform other administrative and clerical duties. They must continually update their skills so that they can instruct and use the latest technology in the classroom.

Teachers often work with students from varied ethnic, racial, and religious backgrounds. With growing minority populations in most parts of the country, it is important for teachers to work effectively with a diverse student population. Accordingly, some schools offer training to help teachers enhance their awareness and understanding of different cultures. Teachers may also include multicultural programming in their lesson plans, to address the needs of all students, regardless of their cultural background. Teachers design classroom presentations to meet students' needs and abilities. They also work with students individually.

Teachers plan, evaluate, and assign lessons; prepare, administer, and grade tests; listen to oral presentations; and maintain classroom discipline. They observe and evaluate a student's performance and potential and increasingly are asked to use new assessment methods. For example, teachers may examine a portfolio of a student's artwork or writing in order to judge the student's overall progress. They then can provide additional assistance in areas in which a student needs help. Teachers also grade papers, prepare report cards, and meet with parents and school staff to discuss a student's academic progress or personal problems.

In addition to conducting classroom activities, teachers oversee study halls and homerooms, supervise extracurricular activities, and accompany students on field trips. They may identify students with physical or mental problems and refer the students to the proper authorities. Secondary school teachers occasionally assist students in choosing courses, colleges, and careers. Teachers also participate in education conferences and workshops. In recent years, site-based management, which allows teachers and parents to participate actively in management decisions regarding school operations, has gained popularity.

In many schools, teachers are increasingly involved in making decisions regarding the budget, personnel, textbooks, curriculum design, and teaching methods.

Working Conditions

Seeing students develop new skills and gain an appreciation of knowledge and learning can be very rewarding. However, teaching may be frustrating when one is dealing with unmotivated or disrespectful students. Occasionally, teachers must cope with unruly behavior and violence in the schools. Teachers may experience stress in dealing with large classes, students from disadvantaged or multicultural backgrounds, or heavy workloads. Inner-city schools in particular, may be run down and lack the amenities of schools in wealthier communities. Accountability standards also may increase stress levels, with teachers expected to produce students who are able to exhibit satisfactory performance on standardized tests in core subjects.

Teachers are sometimes isolated from their colleagues because they work alone in a classroom of students. However, some schools allow teachers to work in teams and with mentors to enhance their professional development. Including school duties performed outside the classroom, many teachers work more than 40 hours a week. Part-time schedules are more common among preschool and kindergarten teachers. Although some school districts have gone to all-day kindergartens, most kindergarten teachers still teach two kindergarten classes a day. Most teachers work the traditional 10-month school year with a 2-month vacation during the summer.

During the vacation break, those on the 10-month schedule may teach in summer sessions, take other jobs, travel, or pursue personal interests. Many enroll in college courses or workshops to continue their education. Teachers in districts with a year-round schedule typically work 8 weeks, are on vacation for 1 week, and have a 5-week midwinter break. Preschool teachers working in daycare settings often work year round. Most States have tenure laws that prevent teachers from being fired without just cause and due process. Teachers may obtain tenure after they have satisfactorily completed a probationary period of teaching, normally 3 years. Tenure does not absolutely guarantee a job, but it does provide some security.

Training, Other Qualifications, and Advancement

All 50 States and the District of Columbia require public school teachers to be licensed. Licensure is not required for teachers in private schools. Usually licensure is granted by the State Board of Education or a licensure advisory committee. Teachers may be licensed to teach the early childhood grades (usually preschool through grade 3); the elementary grades (grades 1 through 6 or 8); the middle grades (grades 5 through 8); a secondary-education subject area (usually grades 7 through 12); or a special subject, such as reading or music (usually grades kindergarten through 12).

Requirements for regular licenses to teach kindergarten through grade 12 vary by State. However, all States require general education teachers to have a bachelor's degree and to have completed an approved teacher training program with a prescribed number of subject and education credits, as well as supervised practice teaching. Some States also require technology training and the attainment of a minimum grade point average. A number of States require that teachers obtain a master's degree in education within a specified period after they begin teaching. Almost all States require applicants for a teacher's license to be tested for competency in basic skills, such as reading and writing, and in teaching.

Almost all also require the teacher to exhibit proficiency in his or her subject. Nowadays, school systems are moving toward implementing performance-based systems for licensure, which usually require the teacher to demonstrate satisfactory teaching performance over an extended period in order to obtain a provisional license, in addition to passing an examination in one's subject. Most States require continuing education for renewal of the teacher's license. Many States have reciprocity agreements that make it easier for teachers licensed in one State to become licensed in another. Many States offer alternative licensure programs for teachers who have bachelor's degrees in the subject they will teach, but who lack the necessary education courses required for a regular license.

Alternative licensure programs originally were designed to ease shortages of teachers of certain subjects, such as mathematics and science. The programs have expanded to attract other people into teaching, including recent college graduates and those changing from another career to teaching. In some programs, individuals begin teaching quickly under provisional licensure. After working under the close supervision of experienced educators for 1 or 2 years while taking education courses outside school hours, they receive regular licensure if they have progressed satisfactorily. In other programs, college graduates who do not meet licensure requirements take only those courses that they lack and then become licensed. This approach may take 1 or 2 semesters of full-time study.

States may issue emergency licenses to individuals who do not meet the requirements for a regular license when schools cannot attract enough qualified teachers to fill positions. Teachers who need to be licensed may enter programs that grant a master's degree in education, as well as a license. In many States, vocational teachers have many of the same requirements for teaching as their academic counterparts. However, because knowledge and experience in a particular field are important criteria for the job, some States will license vocational education teachers without a bachelor's degree, provided they can demonstrate expertise in their field. A minimum number of hours in education courses may also be required.

Licensing requirements for preschool teachers also vary by State. Requirements for public preschool teachers are generally higher than those for private preschool teachers. Some States require a bachelor's degree in early childhood education, others require an associate's degree, and still others require certification by a nationally recognized authority. The Child

Development Associate (CDA) credential, the most common type of certification, requires a mix of classroom training and experience working with children, along with an independent assessment of an individual's competence. In some cases, teachers of kindergarten through high school may attain professional certification in order to demonstrate competency beyond that required for a license.

The National Board for Professional Teaching Standards offers a voluntary national certification. To become nationally accredited, experienced teachers must prove their aptitude by compiling a portfolio showing their work in the classroom and by passing a written assessment and evaluation of their teaching knowledge. Currently, teachers may become certified in a variety of areas, on the basis of the age of the students and, in some cases, the subject taught. For example, teachers may obtain a certificate for teaching English language arts to early adolescents (aged 11 to 15), or they may become certified as early childhood generalists.

All States recognize national certification, and many States and school districts provide special benefits to teachers holding such certification. Benefits typically include higher salaries and reimbursement for continuing education and certification fees. In addition, many States allow nationally certified teachers to carry a license from one State to another. The National Council for Accreditation of Teacher Education currently accredits more than 550 teacher education programs across the United States. Generally, 4-year colleges require students to wait until their sophomore year before applying for admission to teacher education programs.

Traditional education programs for kindergarten and elementary school teachers include courses—designed specifically for those preparing to teach—in mathematics, physical science, social science, music, art, and literature, as well as prescribed professional education courses, such as philosophy of education, psychology of learning, and teaching methods. Aspiring secondary school teachers most often major in the subject they plan to teach while also taking a program of study in teacher preparation. Teacher education programs are now required to include classes in the use of computers and other technologies in order to maintain their accreditation. Most programs require students to perform a student-teaching internship.

Many States now offer professional development schools—partnerships between universities and elementary or secondary schools. Students enter these 1-year programs after completion of their bachelor's degree. Professional development schools merge theory with practice and allow the student to experience a year of teaching firsthand, under professional guidance. In addition to being knowledgeable in their subject, teachers must have the ability to communicate, inspire trust and confidence, and motivate students, as well as understand the students' educational and emotional needs. Teachers must be able to recognize and respond to individual and cultural differences in students and employ different teaching methods that will result in higher student achievement. They should be organized, dependable, patient, and creative.

Teachers also must be able to work cooperatively and communicate effectively with other teachers, support staff, parents, and members of the community. With additional preparation, teachers may move into positions as school librarians, reading specialists, curriculum specialists, or guidance counselors. Teachers may become administrators or supervisors, although the number of these positions is limited and competition can be intense. In some systems, highly qualified, experienced teachers can become senior or mentor teachers, with higher pay and additional responsibilities. They guide and assist less experienced teachers while keeping most of their own teaching responsibilities.

Preschool teachers usually work their way up from assistant teacher, to teacher, to lead teacher—who may be responsible for the instruction of several classes—and, finally, to director of the center. Preschool teachers with a bachelor's degree frequently are qualified to teach kindergarten through grade 3 as well. Teaching at these higher grades often results in higher pay.

Job Outlook

Job opportunities for teachers over the next 10 years will vary from good to excellent, depending on the locality, grade level, and subject taught. Most job openings will be attributable to the expected retirement of a large number of teachers. In addition, relatively high rates of turnover, especially among beginning teachers employed in poor, urban schools, also will lead to numerous job openings for teachers. Competition for qualified teachers among some localities will likely continue, with schools luring teachers from other States and districts with bonuses and higher pay.

Through 2012, overall student enrollments, a key factor in the demand for teachers, are expected to rise more slowly than in the past. As the children of the baby-boom generation get older, smaller numbers of young children will enter school behind them, resulting in average employment growth for all teachers, from preschool through secondary grades. Projected enrollments will vary by region. Fast-growing States in the South and West—particularly California, Texas, Georgia, Idaho, Hawaii, Alaska, and New Mexico—will experience the largest enrollment increases. Enrollments in the Northeast and Midwest are expected to hold relatively steady or decline. The job market for teachers also continues to vary by school location and by subject taught.

Many inner cities—often characterized by overcrowded, ill-equipped schools and higher-than-average poverty rates—and rural areas—characterized by their remote location and relatively low salaries—have difficulty attracting and retaining enough teachers, so job prospects should be better in these areas than in suburban districts. Currently, many school districts have difficulty hiring qualified teachers in some subject areas—

mathematics, science (especially chemistry and physics), bilingual education, and foreign languages. Qualified vocational teachers, at both the middle school and secondary school levels, also are currently in demand in a variety of fields.

Specialties that have an adequate number of qualified teachers include general elementary education, physical education, and social studies. Teachers who are geographically mobile and who obtain licensure in more than one subject should have a distinct advantage in finding a job. Increasing enrollments of minorities, coupled with a shortage of minority teachers, should cause efforts to recruit minority teachers to intensify. Also, the number of non-English-speaking students has grown dramatically, creating demand for bilingual teachers and for those who teach English as a second language. The number of teachers employed is dependent as well on State and local expenditures for education and on the enactment of legislation to increase the quality of education.

A number of initiatives, such as reduced class size (primarily in the early elementary grades), mandatory preschool for 4-year-olds, and all-day kindergarten, have been implemented in a few States, but not nationwide. Additional teachers—particularly preschool and early elementary school teachers—will be needed if States or localities implement any of these measures. At the Federal level, legislation that is likely to affect teachers recently was put into place with the enactment of the No Child Left Behind Act. Although the full impact of this act is not yet known, its emphasis on ensuring that all schools hire and retain only qualified teachers, may lead to an increase in funding for schools that currently lack such teachers.

The supply of teachers is expected to increase in response to reports of improved job prospects, better pay, more teacher involvement in school policy, and greater public interest in education. In recent years, the total number of bachelor's and master's degrees granted in education has increased steadily. Because of a shortage of teachers in certain locations, and in anticipation of the loss of a number of teachers to retirement, many States have implemented policies that will encourage more students to become teachers. In addition, more teachers may be drawn from a reserve pool of career changers, substitute teachers, and teachers completing alternative certification programs.

TEACHERS—POSTSECONDARY

Postsecondary teachers instruct students in a wide variety of academic and vocational subjects beyond the high school level that may lead to a degree or simply to improvement in one's knowledge or skills. These teachers include college and university faculty, postsecondary career and technical education teachers, and graduate teaching assistants. College and university faculty make up the majority of postsecondary teachers. They teach and advise more than 15 million full- and part-time college students and perform a significant part of our Nation's research. Faculty also keep up with new developments in their field and may consult with government, business, nonprofit, and community organizations.

Faculty usually are organized into departments or divisions, based on academic subject or field. They usually teach several different related courses in their subject—algebra, calculus, and statistics, for example. They may instruct undergraduate or graduate students, or both. College and university faculty may give lectures to several hundred students in large halls, lead small seminars, or supervise students in laboratories. They prepare lectures, exercises, and laboratory experiments; grade exams and papers; and advise and work with students individually. In universities, they also supervise graduate students' teaching and research.

College faculty work with an increasingly varied student population made up of growing shares of part-time, older, and culturally and racially diverse students. Faculty keep abreast of developments in their field by reading current literature, talking with colleagues, and participating in professional conferences. They may also do their own research to expand knowledge in their field. They may perform experiments; collect and analyze data; and examine original documents, literature, and other source material. From this process, they arrive at conclusions, and publish their findings in scholarly journals, books, and electronic media.

Most college and university faculty extensively use computer technology, including the Internet; electronic mail; software programs, such as statistical packages; and CD-ROMs. They may use computers in the classroom as teaching aids and may post course content, class notes, class schedules, and other information on the Internet. Some faculty are increasingly using sophisticated telecommunications and videoconferencing equipment and the Internet to teach courses to students at remote sites. The use of e-mail, chat rooms, and other techniques has greatly improved communications between students and teachers and among students.

Most faculty members serve on academic or administrative committees that deal with the policies of their institution, departmental matters, academic issues, curricula, budgets, equipment purchases, and hiring. Some work with student and community organizations. Department chairpersons are faculty members who usually teach some courses but have heavier administrative responsibilities. The proportion of time spent on research, teaching, administrative, and other duties varies by individual circumstance and type of institution. Faculty members at universities normally spend a significant part of their time doing research; those in 4-year colleges, somewhat less; and those in 2-year colleges, relatively little.

The teaching load, however, often is heavier in 2-year colleges and somewhat lighter at 4-year institutions. Full professors at all types of institutions usually spend a larger portion of their time conducting research than do assistant professors, instructors, and lecturers. Postsecondary vocational education teachers, also known as postsecondary career and technical education teachers, provide instruction for occupations that require specialized training, but may not require a 4-year degree, such as welder, dental hygienist, x-ray technician, auto mechanic, and cosmetologist. Classes often are taught in an industrial or laboratory setting where students are provided hands-on experience. For example, welding instructors show students various welding techniques and essential safety practices, watch them use tools and equipment, and have them repeat procedures until they meet the specific standards required by the trade.

Increasingly, career and technical education teachers are integrating academic and vocational curriculums so that students obtain a variety of skills that can be applied to the "real world." Career and technical education teachers have many of the same responsibilities that other college and university faculty have. They must prepare lessons, grade papers, attend faculty meetings, and keep abreast of developments in their field. Career and technical education teachers at community colleges and career and technical schools also often play a key role in students' transition from school to work by helping to establish internship programs for students and by providing information about prospective employers.

Graduate teaching assistants, often referred to as graduate TAs, assist faculty, department chairs, or other professional staff at colleges and universities by performing teaching or teaching-related duties. In addition to their work responsibilities, assistants have their own school commitments, as they are also students who are working towards earning a graduate degree, such as a Ph.D. Some teaching assistants have full responsibility for teaching a course—usually one that is introductory in nature—which can include preparation of lectures and exams, and assigning final grades to students. Others provide assistance to faculty members, which may consist of a variety of tasks such as grading papers, monitoring exams, holding office hours or help-sessions for students, conducting laboratory sessions, or administering quizzes to the class. Teaching assistants generally meet initially with the faculty member whom they are going to assist in order to determine exactly what is expected of them, as each faculty member may have his or her own needs. For example, some faculty members prefer assistants to sit in on classes, while others assign them other tasks to do during class time. Graduate teaching assistants may work one-on-one with a faculty member or, for large classes, they may be one of several assistants.

Working Conditions

Postsecondary teachers usually have flexible schedules. They must be present for classes, usually 12 to 16 hours per week, and for faculty and committee meetings. Most establish regular office hours for student consultations, usually 3 to 6 hours per week. Otherwise, teachers are free to decide when and where they will work, and how much time to devote to course preparation, grading, study, research, graduate student supervision, and other activities. Some teach night and weekend classes. This is particularly true for teachers at 2-year community colleges or institutions with large enrollments of older students who have full-time jobs or family responsibilities.

Most colleges and universities require teachers to work 9 months of the year, which allows them the time to teach additional courses, do research, travel, or pursue nonacademic interests during the summer and school holidays. Colleges and universities usually have funds to support research or other professional development needs, including travel to conferences and research sites. About 3 out of 10 college and university faculty worked part time in 2002. Some part-timers, known as "adjunct faculty," have primary jobs outside of academia—in government, private industry, or nonprofit research—and teach "on the side." Others prefer to work part-time hours or seek full-time jobs but are unable to obtain them due to intense competition for available openings. Some work part time in more than one institution.

Many adjunct faculty are not qualified for tenure-track positions because they lack a doctoral degree. University faculty may experience a conflict between their responsibilities to teach students and the pressure to do research and publish their findings. This may be a particular problem for young faculty seeking advancement in 4-year research universities. Also, recent cutbacks and the hiring of more part-time faculty have put a greater administrative burden on full-time faculty. Requirements to teach online classes also have added greatly to the workloads of postsecondary teachers. Many find that developing the courses to put online, plus learning how to operate the technology and answering large amounts of e-mail, is very time-consuming.

Like college and university faculty, there is usually a great deal of flexibility in graduate TAs' work schedules, which allows them the time to pursue their own academic coursework and studies. The number of hours that TAs work varies depending on their assignments. Work may be stressful, particularly when assistants are given full responsibility for teaching a class; however, these types of positions allow graduate students the opportunity to gain valuable teaching experience. This experience is especially helpful for those graduate teaching assistants who seek to become faculty members at colleges and universities after completing their degree.

Training, Other Qualifications, and Advancement

The education and training required of postsecondary teachers varies widely, depending on the subject taught and educational institution employing them. Educational requirements for teachers are generally the highest at 4-year research universities but, at career and technical institutes, experience and expertise in a related occupation is the most valuable qualification. Postsecondary teachers should communicate and relate well with students, enjoy working with them, and be able to motivate them. They should have inquiring and analytical minds, and a strong desire to pursue and disseminate knowledge. Additionally, they must be self-motivated and able to work in an environment in which they receive little direct supervision.

Training requirements for postsecondary career and technical education teachers vary by State and by subject. In general, teachers need a bachelor's or higher degree, plus work or other experience in their field. In some fields, a license or certificate that demonstrates one's qualifications may be all that is required. Teachers update their skills through continuing education, in order to maintain certification. They must also maintain ongoing dialogue with businesses to determine the most current skills needed in the workplace. Four-year colleges and universities usually consider doctoral degree holders for full-time, tenure-track positions, but may hire master's degree holders or doctoral candidates for certain disciplines, such as the arts, or for part-time and temporary jobs. Most college and university faculty are in four academic ranks—professor, associate professor, assistant professor, and instructor. These positions usually are considered to be tenure-track positions.

Most faculty members are hired as instructors or assistant professors. A smaller number of additional faculty members, called lecturers, are usually employed on contracts for a single academic term and are not on the tenure track. In 2-year colleges, master's degree holders fill most full-time positions. However, with increasing competition for available jobs, institutions can be more selective in their hiring practices. Many 2-year institutions increasingly prefer job applicants to have some teaching experience or experience with distance learning. Preference also may be given to those holding dual master's degrees, because they can teach more subjects. In addition, with greater competition for jobs, master's degree holders may find it increasingly difficult to obtain employment as they are passed over in favor of candidates holding a Ph.D.

Doctoral programs take an average of 6 to 8 years of full-time study beyond the bachelor's degree, including time spent completing a master's degree and a dissertation. Some programs, such as those in the humanities, take longer to complete; others, such as those in engineering, usually are shorter. Candidates specialize in a subfield of a discipline—for example, organic chemistry, counseling psychology, or European history—but also take courses covering the entire discipline. Programs include 20 or more increasingly specialized courses and seminars plus comprehensive examinations on all major areas of the field.

Candidates also must complete a dissertation—a written report on original research in the candidate's major field of study. The dissertation sets forth an original hypothesis or proposes a model and tests it. Students in the natural sciences and engineering usually do laboratory work; in the humanities, they study original documents and other published material. The dissertation is done under the guidance of one or more faculty advisors and usually takes 1 or 2 years of full-time work. In some fields, particularly the natural sciences, some students spend an additional 2 years on postdoctoral research and study before taking a faculty position. Some Ph.D.s extend postdoctoral appointments, or take new ones, if they are unable to find a faculty job. Most of these appointments offer a nominal salary.

Obtaining a position as a graduate teaching assistant is a good way to gain college teaching experience. To qualify, candidates must be enrolled in a graduate school program. In addition, some colleges and universities require teaching assistants to attend classes or take some training prior to being given responsibility for a course. Although graduate teaching assistants usually work at the institution and in the department where they are earning their degree, teaching or internship positions for graduate students at institutions that do not grant a graduate degree have become more common in recent years. For example, a program called Preparing Future Faculty, administered by the Association of American Colleges and Universities and the Council of Graduate Schools, has led to the creation of many now-independent programs that offer graduate students at research universities the opportunity to work as teaching assistants at other types of institutions, such as liberal arts or community colleges.

Working with a mentor, the graduate students teach classes and learn how to improve their teaching techniques. They may attend faculty and committee meetings, develop a curriculum, and learn how to balance the teaching, research, and administrative roles that faculty play. These programs provide valuable learning opportunities for graduate students interested in teaching at the postsecondary level, and also help to make these students aware of the differences among the various types of institutions at which they may someday work. For faculty, a major step in the traditional academic career is attaining tenure. New tenure-track faculty usually are hired as instructors or assistant professors, and must serve a period—usually 7 years—under term contracts.

At the end of the period, their record of teaching, research, and overall contribution to the institution is reviewed; tenure is granted if the review is favorable. Those denied tenure usually must leave the institution. Tenured professors cannot be fired without just cause and due process. Tenure protects the faculty's academic freedom—the ability to teach and conduct research without fear of being fired for advocating unpopular ideas. It also gives both faculty and institutions the stability needed for effective research and teaching, and provides financial security for faculty. Some institutions have adopted post-tenure review policies to encourage ongoing evaluation of tenured faculty. The number of tenure-track positions is expected to decline as institutions seek flexibility in dealing with financial matters and changing student interests.

Institutions will rely more heavily on limited term contracts and part-time, or adjunct, faculty, thus shrinking the total pool of tenured faculty. In a trend that is expected to continue, some institutions now offer limited-term contracts to prospective faculty—typically 2-, 3-, or 5-year, full-time contracts. These contracts may be terminated or extended when they expire.

Institutions are not obligated to grant tenure to the contract holders. In addition, some institutions have limited the percentage of faculty who can be tenured. For most postsecondary teachers, advancement involves a move into administrative and managerial positions, such as departmental chairperson, dean, and president.

At 4-year institutions, such advancement requires a doctoral degree. At 2-year colleges, a doctorate is helpful but not usually required, except for advancement to some top administrative positions.

TEACHER ASSISTANTS

Teacher assistants provide instructional and clerical support for classroom teachers, allowing teachers more time for lesson planning and teaching. Teacher assistants tutor and assist children in learning class material using the teacher's lesson plans, providing students with individualized attention. Teacher assistants also supervise students in the cafeteria, schoolyard, and hallways, or on field trips. They record grades, set up equipment, and help prepare materials for instruction. Teacher assistants also are called teacher aides or instructional aides. Some assistants refer to themselves as paraeducators or paraprofessionals.

Some teacher assistants perform exclusively noninstructional or clerical tasks, such as monitoring nonacademic settings. Playground and lunchroom attendants are examples of such assistants. Most teacher assistants, however, perform a combination of instructional and clerical duties. They generally provide instructional reinforcement to children, under the direction and guidance of teachers. They work with students individually or in small groups—listening while students read, reviewing or reinforcing class lessons, or helping them find information for reports. At the secondary school level, teacher assistants often specialize in a certain subject, such as math or science.

Teacher assistants often take charge of special projects and prepare equipment or exhibits, such as for a science demonstration. Some assistants work in computer laboratories, helping students using computers and educational software programs. In addition to instructing, assisting, and supervising students, teacher assistants grade tests and papers, check homework, keep health and attendance records, do typing and filing, and duplicate materials. They also stock supplies, operate audiovisual equipment, and keep classroom equipment in order.

Many teacher assistants work extensively with special education students. As schools become more inclusive, integrating special education students into general education classrooms, teacher assistants in general education and special education classrooms increasingly assist students with disabilities. Teacher assistants attend to a disabled student's physical needs, including feeding, teaching good grooming habits, or assisting students riding the schoolbus. They also provide personal attention to students with other special needs, such as those from disadvantaged families, those who speak English as a second language, or those who need remedial education.

Teacher assistants help assess a student's progress by observing performance and recording relevant data. Teacher assistants also work with infants and toddlers who have developmental delays or other disabilities. Under the guidance of a teacher or therapist, teacher assistants perform exercises or play games to help the child develop physically and behaviorally. Some teacher assistants work with young adults to help them obtain a job or to apply for community services for the disabled.

Working Conditions

Approximately 4 in 10 teacher assistants work part time. However, even among full-time workers, nearly 40 percent work less than 8 hours per day. Most assistants who provide educational instruction work the traditional 9- to 10-month school year. Teacher assistants work in a variety of settings—including private homes and preschools, and local government offices, where they would deal with young adults—but most work in classrooms in elementary, middle, and secondary schools. They also work outdoors supervising recess when weather allows, and they spend much of their time standing, walking, or kneeling.

Seeing students develop and gain appreciation of the joy of learning can be very rewarding. However, working closely with students can be both physically and emotionally tiring. Teacher assistants who work with special education students often perform more strenuous tasks, including lifting, as they help students with their daily routine. Those who perform clerical work may tire of administrative duties, such as copying materials or typing.

Training, Other Qualifications, and Advancement

Educational requirements for teacher assistants vary by State or school district and range from a high school diploma to some college training, although employers increasingly prefer applicants with some college training. Teacher assistants with instructional responsibilities usually require more training than do those who do not perform teaching tasks. In addition, as a result of the No Child Left Behind Act of 2001, teacher assistants in Title 1 schools—those with a large proportion of students from low-income households—will be required to meet one of three requirements: have a minimum of 2 years of college, hold a 2-year or higher degree, or pass a rigorous state and local assessment. Many schools also require previous experience in working with children and a valid driver's license. Some schools may require the applicant to pass a background check.

A number of 2-year and community colleges offer associate degree programs that prepare graduates to work as teacher assistants. However, most teacher assistants receive on-the-job training. Those who tutor and review lessons with students must have a thorough understanding of class materials and instructional methods, and should be familiar with the organization and operation of a school. Teacher assistants also must know how to operate audiovisual equipment, keep records, and prepare instructional materials, as well as have adequate computer skills. Teacher assistants should enjoy working with children from a wide range of cultural backgrounds, and be able to handle classroom situations with fairness and patience.

Teacher assistants also must demonstrate initiative and a willingness to follow a teacher's directions. They must have good writing skills and be able to communicate effectively with students and teachers. Teacher assistants who speak a second language, especially Spanish, are in great demand for communicating with growing numbers of students and parents whose primary language is not English. Advancement for teacher assistants—usually in the form of higher earnings or increased responsibility—comes primarily with experience or additional education. Some school districts provide time away from the job or tuition reimbursement so that teacher assistants can earn their bachelor's degrees and pursue licensed teaching positions. In return for tuition reimbursement, assistants are often required to teach a certain length of time for the school district.

Job Outlook

Employment of teacher assistants is expected to grow somewhat faster than the average for all occupations through 2012. Although school enrollments are projected to increase only slowly over the next decade, the student population for which teacher assistants are most needed—special education students and students for whom English is not their first language—is expected to increase more rapidly than the general school-age population. Legislation that requires students with disabilities and non-native English speakers to receive an education "equal" to that of other students, will generate jobs for teacher assistants to accommodate these students' special needs. Children with special needs require much personal attention, and special education teachers, as well as general education teachers with special education students, rely heavily on teacher assistants.

Additionally, a greater focus on educational quality and accountability, as required by the No Child Left Behind Act, is likely to lead to an increased demand for teacher assistants. Growing numbers of teacher assistants will be needed to help teachers prepare students for standardized testing and to provide extra assistance to students who perform poorly on standardized tests. An increasing number of afterschool programs and summer programs also will create new opportunities for teacher assistants. In addition to those stemming from employment growth, numerous job openings will arise as assistants transfer to other occupations or leave the labor force to assume family responsibilities, to return to school, or for other reasons characteristic of occupations that require limited formal education and offer relatively low pay. Opportunities for teacher assistant jobs are expected to be best for persons with at least 2 years of formal education after high school.

Persons who can speak a foreign language should be in particular demand in school systems with large numbers of students whose families do not speak English at home. Demand is expected to vary by region of the country. Areas in which the population and school enrollments are expanding rapidly, such as many communities in the South and West, should have rapid growth in the demand for teacher assistants.

COUNSELORS

Counselors assist people with personal, family, educational, mental health, and career decisions and problems. Their duties depend on the individuals they serve and on the settings in which they work. Educational, vocational, and school counselors provide individuals and groups with career and educational counseling. In school settings—elementary through postsecondary—they are usually called school counselors and they work with students, including those considered to be at risk and those with special needs. They advocate for students and work with other individuals and organizations to promote the academic, career, and personal and social development of children and youths.

School counselors help students evaluate their abilities, interests, talents, and personality characteristics in order to develop realistic academic and career goals. Counselors use interviews, counseling sessions, tests, or other methods in evaluating and advising students. They also operate career information centers and career education programs. High school counselors advise students regarding college majors, admission requirements, entrance exams, financial aid, trade or technical schools, and apprenticeship programs. They help students develop job search skills such as resume writing and interviewing techniques. College career planning and placement counselors assist alumni or students with career development and job-hunting techniques.

Elementary school counselors observe younger children during classroom and play activities and confer with their teachers and parents to evaluate the children's strengths, problems, or special needs. They also help students develop good study habits. Elementary school counselors do less vocational and academic counseling than do secondary school counselors. School counselors at all levels help students understand and deal with social, behavioral, and personal problems. These counselors emphasize preventive and developmental counseling to provide students with the life skills needed to deal with problems before they occur and to enhance the student's personal, social, and academic growth.

Counselors provide special services, including alcohol and drug prevention programs and conflict resolution classes. Counselors also try to identify cases of domestic abuse and other family problems that can affect a student's development. Counselors work with students individually, with small groups, or with entire classes. They consult and collaborate with parents, teachers, school administrators, school psychologists, medical professionals, and social workers in order to develop and implement strategies to help students be successful in the education system. Vocational counselors who provide mainly career counseling outside the school setting are also referred to as employment counselors or career counselors. Their chief focus is helping individuals with their career decisions.

Vocational counselors explore and evaluate the client's education, training, work history, interests, skills, and personality traits, and arrange for aptitude and achievement tests to assist in making career decisions. They also work with individuals to develop their job search skills, and they assist clients in locating and applying for jobs. In addition, career counselors provide support to persons experiencing job loss, job stress, or other career transition issues. Rehabilitation counselors help people deal with the personal, social, and vocational effects of disabilities. They counsel people with disabilities resulting from birth defects, illness or disease, accidents, or the stress of daily life. They evaluate the strengths and limitations of individuals, provide personal and vocational counseling, and arrange for medical care, vocational training, and job placement.

Rehabilitation counselors interview both individuals with disabilities and their families, evaluate school and medical reports, and confer and plan with physicians, psychologists, occupational therapists, and employers to determine the capabilities and skills of the individual. Conferring with the client, they develop a rehabilitation program that often includes training to help the person develop job skills. Rehabilitation counselors also work toward increasing the client's capacity to live independently. Mental health counselors work with individuals, families, and groups to address and treat mental and emotional disorders and to promote optimum mental health. They are trained in a variety of therapeutic techniques used to address a wide range of issues, including depression, addiction and substance abuse, suicidal impulses, stress management, problems with self-esteem, issues associated with aging, job and career concerns, educational decisions, issues related to mental and emotional health, and family, parenting, and marital or other relationship problems.

Mental health counselors often work closely with other mental health specialists, such as psychiatrists, psychologists, clinical social workers, psychiatric nurses, and school counselors. (Information on other mental health specialists appears in the Handbook statements on physicians and surgeons, psychologists, registered nurses, and social workers.) Substance abuse and behavioral disorder counselors help people who have problems with alcohol, drugs, gambling, and eating disorders. They counsel individuals who are addicted to drugs, helping them identify behaviors and problems related to their addiction. These counselors hold sessions for one person, for families, or for groups of people.

Marriage and family therapists apply principles, methods, and therapeutic techniques to individuals, family groups, couples, or organizations for the purpose of resolving emotional conflicts. In doing so, they modify people's perceptions and behaviors, enhance communication and understanding among all family members, and help to prevent family and individual crises. Marriage and family therapists also may engage in psychotherapy of a nonmedical nature, with appropriate referrals to psychiatric resources, and in research and teaching in the overall field of human development and interpersonal relationships.

Other counseling specialties include gerontological, multicultural, and genetic counseling. A gerontological counselor provides services to elderly persons who face changing lifestyles because of health problems; the counselor helps families cope with the changes. A multicultural counselor helps employers adjust to an increasingly diverse workforce. Genetic counselors provide information and support to families who have members with birth defects or genetic disorders and to families who may be at risk for a variety of inherited conditions. These counselors identify families at risk, investigate the problem that is present in the family, interpret information about the disorder, analyze inheritance patterns and risks of recurrence, and review available options with the family.

Working Conditions

Most school counselors work the traditional 9- to 10-month school year with a 2- to 3-month vacation, although increasing numbers are employed on 10½- or 11-month contracts. They usually work the same hours that teachers do. College career planning and placement counselors work long and irregular hours during student recruiting periods. Rehabilitation counselors usually work a standard 40-hour week. Self-employed counselors and those working in mental health and community agencies, such as substance abuse and behavioral disorder counselors, frequently work evenings to counsel clients who work during the day.

Both mental health counselors and marriage and family therapists also often work flexible hours, to accommodate families in crisis or working couples who must have evening or weekend appointments. Counselors must possess high physical and emotional energy to handle the array of problems they address. Dealing daily with these problems can cause stress. Because privacy is essential for confidential and frank discussions with clients, counselors usually have private offices.

Training, Other Qualifications, and Advancement

All States require school counselors to hold State school counseling certification and to have completed at least some graduate course work; most require the completion of a master's degree. Some States require public school counselors to have both counseling and teaching certificates and to have had some teaching experience before receiving certification. For counselors based outside of schools, 47 States and the District of Columbia had some form of counselor credentialing, licensure, certification, or registration that governed their practice of counseling. Requirements typically include the completion of a master's degree in counseling, the accumulation of 2 years or 3,000 hours of supervised clinical experience beyond the master's degree level, the passage of a State-recognized exam, adherence to ethical codes and standards, and the satisfaction of annual continuing education requirements.

Counselors must be aware of educational and training requirements that are often very detailed and that vary by area and by counseling specialty. Prospective counselors should check with State and local governments, employers, and national voluntary certification organizations in order to determine which requirements apply. As mentioned, a master's degree is typically required to be licensed or certified as a counselor. A bachelor's degree often qualifies a person to work as a counseling aide, rehabilitation aide, or social service worker. Some States require counselors in public employment to have a master's degree; others accept a bachelor's degree with appropriate counseling courses.

Counselor education programs in colleges and universities usually are in departments of education or psychology. Fields of study include college student affairs, elementary or secondary school counseling, education, gerontological counseling, marriage and family counseling, substance abuse counseling, rehabilitation counseling, agency or community counseling, clinical mental health counseling, counseling psychology, career counseling, and related fields. Courses are grouped into eight core areas: Human growth and development, social and cultural diversity, relationships, group work, career development, assessment, research and program evaluation, and professional identity.

In an accredited master's degree program, 48 to 60 semester hours of graduate study, including a period of supervised clinical experience in counseling, are required for a master's degree. In 2003, 176 institutions offered programs in counselor education—including career, community, gerontological, mental health, school, student affairs, and marriage and family counseling—that were accredited by the Council for Accreditation of Counseling and Related Educational Programs (CACREP). CACREP also recognizes many counselor education programs, apart from those in the 176 accredited institutions, that use alternative instruction methods, such as distance learning.

Programs that use such alternative instruction methods are evaluated on the basis of the same standards for accreditation that CACREP applies to programs that employ the more traditional methods. Another organization, the Council on Rehabilitation Education (CORE), accredits graduate programs in rehabilitation counseling. Accredited master's degree programs include a minimum of 2 years of full-time study, including 600 hours of supervised clinical internship experience. Many counselors elect to be nationally certified by the National Board for Certified Counselors, Inc. (NBCC), which grants the general practice credential "National Certified Counselor." To be certified, a counselor must hold a master's or higher degree, with a concentration in counseling, from a regionally accredited college or university; must have at least 2 years of supervised field experience in a counseling setting (graduates from counselor education programs accredited by CACREP are exempted); must provide two professional endorsements, one of which must be from a recent supervisor; and must have a passing score on the NBCC's National Counselor Examination for Licensure and Certification (NCE).

This national certification is voluntary and is distinct from State certification. However, in some States, those who pass the national exam are exempted from taking a State certification exam. NBCC also offers specialty certification in school, clinical mental health, and addiction counseling. Beginning January 1, 2004, new candidates for NBCC's National Certified School counselor (NCSC) credential must pass a practical simulation examination in addition to fulfilling the current requirements. To maintain their certification, counselors retake and pass the NCE or complete 100 hours of acceptable continuing education credit every 5 years. Another organization, the Commission on Rehabilitation Counselor Certification, offers voluntary national certification for rehabilitation counselors.

Many employers require rehabilitation counselors to be nationally certified. To become certified, rehabilitation counselors usually must graduate from an accredited educational program, complete an internship, and pass a written examination. (Certification requirements vary according to an applicant's educational history. Employment experience, for example, is required for those with a counseling degree in a specialty other than rehabilitation.) After meeting these requirements, candidates are designated "Certified Rehabilitation Counselors." To maintain their certification, counselors must successfully retake the certification exam or complete 100 hours of acceptable continuing education credit every 5 years. Other counseling organizations also offer certification in particular counseling specialties. Usually these are voluntary, but having one may enhance one's job prospects.

Some employers provide training for newly hired counselors. Others may offer time off or provide help with tuition if it is needed to complete a graduate degree. Counselors must participate in graduate studies, workshops, and personal studies to maintain their certificates and licenses. Persons interested in counseling should have a strong interest in helping others and should possess the ability to inspire respect, trust, and confidence. They should be able to work independently or as part of a team. Counselors must follow the code of ethics associated with their respective certifications and licenses. Prospects for advancement vary by counseling field. School counselors can move to a larger school; become directors or supervisors of counseling, guidance, or pupil personnel services; or, usually with further graduate education, become counselor educators, counseling psychologists, or school administrators.

Some counselors choose to work for a State's department of education. For marriage and family therapists, doctoral education in family therapy emphasizes the training of supervisors, teachers, researchers, and clinicians in the discipline Counselors can become supervisors or administrators in their agencies. Some counselors move into research, consulting, or college teaching or go into private or group practice.

Job Outlook

Overall employment of counselors is expected to grow faster than the average for all occupations through 2012, and job opportunities should be very good because there are usually more job openings than graduates of counseling programs. In addition, numerous job openings will occur as many counselors retire or leave the profession. Employment of educational, vocational, and school counselors is expected to grow as fast as the average for all occupations as a result of: increasing student enrollments, particularly in secondary and postsecondary schools; State legislation requiring counselors in elementary schools; and an expansion in the responsibilities of counselors. For example, counselors are becoming more involved in crisis and preventive counseling, helping students deal with issues ranging from drug and alcohol abuse to death and suicide.

Although schools and governments realize the value of counselors in achieving academic success in their students, budget constraints at every school level will dampen job growth of school counselors. However, Federal grants and subsidies may fill in the gaps and allow the current ongoing reduction in student-to-counselor ratios to continue. Demand for vocational or career counselors should grow as the notion of staying in one job over a lifetime continues to be rejected and replaced by the concept of managing one's own career and taking responsibility for it. In addition, changes in welfare laws that require beneficiaries to work will continue to create demand for counselors by State and local governments.

Other opportunities for employment counselors will arise in private job-training centers that provide training and other services to laid-off workers, as well as to those seeking a new or second career or wanting to upgrade their skills. Demand is expected to be strong for substance abuse and behavioral, mental health, and marriage and family therapists and for rehabilitation counselors, for a variety of reasons. For one, California and a few other States have recently passed laws requiring substance abuse treatment instead of jail for people caught possessing a drug. This shift will require more substance abuse counselors in those States. Second, the increasing availability of funds to build statewide networks to improve services for children and adolescents with serious emotional disturbances and for their family members should increase employment opportunities for counselors.

Under managed care systems, insurance companies are increasingly providing for reimbursement of counselors as a less costly alternative to psychiatrists and psychologists. Also, legislation is pending that may provide counseling services to Medicare recipients. The number of people who will need rehabilitation counseling is expected to grow as the population continues to age and as advances in medical technology continue to save lives that only a few years ago would have been lost. In addition, legislation requiring equal employment rights for people with disabilities will spur demand for counselors, who not only will help these people make a transition into the workforce, but also will help companies comply with the law.

Employment of mental health counselors and marriage and family therapists will grow as the Nation becomes more comfortable seeking professional help for a variety of health and personal and family problems. Employers also are increasingly offering employee assistance programs that provide mental health and alcohol and drug abuse services. More people are expected to use these services as society focuses on ways of developing mental well-being, such as controlling stress associated with job and family responsibilities.

CHILDCARE WORKERS

Childcare workers nurture and teach children of all ages in childcare centers, nursery schools, preschools, public schools, private households, family childcare homes, and before- and afterschool programs. These workers play an important role in a child's development by caring for the child when parents are at work or away for other reasons. Some parents enroll their children in nursery schools or childcare centers primarily to provide them with the opportunity to interact with other children. In addition to attending to children's basic needs, these workers organize activities that stimulate the children's physical, emotional, intellectual, and social growth. They help children to explore their interests, develop their talents and independence, build self-esteem, and learn how to get along with others.

Private household workers who are employed on an hourly basis usually are called *babysitters*. These childcare workers bathe, dress, and feed children; supervise their play; wash their clothes; and clean their rooms. They also may put them to bed and waken them, read to them, involve them in educational games, take them for doctors' visits, and discipline them. Those who are in charge of infants, sometimes called infant nurses, also prepare bottles and change diapers.

Nannies generally take care of children from birth to age 10 or 12, tending to the child's early education, nutrition, health, and other needs. They also may perform the duties of a general housekeeper, including general cleaning and laundry duties. Childcare workers spend most of their day working with children. However, they do maintain contact with parents or guardians through informal meetings or scheduled conferences to discuss each child's progress and needs. Many childcare workers keep records of each child's progress and suggest ways in which parents can stimulate their child's learning and development at home. Some preschools, childcare centers, and before- and after-school programs actively recruit parent volunteers to work with the children and participate in administrative decisions and program planning.

Most childcare workers perform a combination of basic care and teaching duties. Through many basic care activities, childcare workers provide opportunities for children to learn. For example, a worker who shows a child how to tie a shoelace teaches the child while also providing for that child's basic care needs. Childcare programs help children to learn about trust and to gain a sense of security. Young children learn mainly through play. Recognizing the importance of play, childcare workers build their program around it. They capitalize on children's play to further language development (storytelling and acting games), improve social skills (working together to build a neighborhood in a sandbox), and introduce scientific and mathematical concepts (balancing and counting blocks when building a bridge or mixing colors when painting). Thus, a less structured approach is used to teach preschool children, including small-group lessons, one-on-one instruction, and learning through creative activities, such as art, dance, and music.

Interaction with peers is an important part of a child's early development. Preschool children in childcare centers have an opportunity to engage in conversation and discussions, and to learn to play and work cooperatively with their classmates. Childcare workers play a vital role in preparing children to build the skills they will need in school. Childcare workers in preschools greet young children as they arrive, help them to remove outer garments, and select an activity of interest. When caring for infants, they feed and change them. To ensure a well-balanced program, childcare workers prepare daily and long-term schedules of activities. Each day's activities balance individual and group play, and quiet and active time. Children are given some freedom to participate in activities in which they are interested.

Concern over school-age children being home alone before and after school has spurred many parents to seek alternative ways for their children to constructively spend their time. The purpose of before- and afterschool programs is to watch over school-age children during the gap between school hours and their parents' work hours. These programs also may operate during the summer and on weekends. Workers in before- and after-school programs may help students with their homework or engage them in other extracurricular activities. These activities may include field trips, learning about computers, painting, photography, and participating in sports.

Some childcare workers may be responsible for taking children to school in the morning and picking them up from school in the afternoon. Before- and afterschool programs may be operated by public school systems, local community centers, or other private organizations. Helping to keep young children healthy is an important part of the job. Childcare workers serve nutritious meals and snacks and teach good eating habits and personal hygiene. They ensure that children have proper rest periods. They identify children who may not feel well or who show signs of emotional or developmental problems and discuss these matters with their supervisor and the child's parents. In some cases, childcare workers help parents to locate programs that will provide basic health services. Early identification of children with special needs—such as those with behavioral, emotional, physical, or learning disabilities—is important to improve their future learning ability. Special education teachers often work with these preschool children to provide the individual attention they need.

Working Conditions

Preschool or childcare facilities include private homes, schools, religious institutions, workplaces in which employers provide care for employees' children, and private buildings. Individuals who provide care in their own homes generally are called family childcare providers. Nannies and babysitters usually work in the pleasant and comfortable homes or apartments of their employers. Most are day workers who live in their own homes and travel to work. Some live in the home of their employer, generally with their own room and bath. They often become part of their employer's family, and may derive satisfaction from caring for them.

Watching children grow, learn, and gain new skills can be very rewarding. While working with children, childcare workers often improve the child's communication, learning, and other personal skills. The work is sometimes routine; however, new activities and challenges mark each day. Childcare can be physically and emotionally taxing, as workers constantly stand, walk, bend, stoop, and lift to attend to each child's interests and problems. To ensure that children receive proper supervision, State or local regulations may require a certain ratio of workers to children. The ratio varies with the age of the children. Child development experts generally recommend that a single caregiver be responsible for no more than 3 or 4 infants (less than 1 year old), 5 or 6 toddlers (1 to 2 years old), or 10 preschool-age children (between 2 and 5 years old).

In before- and afterschool programs, workers may be responsible for many school-age children at one time. The work hours of childcare workers vary widely. Childcare centers usually are open year round, with long hours so that parents can drop off and pick up their children before and after work. Some centers employ full-time and part-time staff with staggered shifts to cover the entire day. Some workers are unable to take regular breaks during the day due to limited staffing. Public and many private preschool programs operate during the typical 9- or 10-month school year, employing both full-time and part-time workers. Family childcare providers have flexible hours and daily routines, but may work long or unusual hours to fit parents' work schedules. Live-in nannies usually work longer hours than do those who have their own homes.

However, if they work evenings or weekends, they may get other time off. Replacement needs in this occupation are high. Many childcare workers leave the occupation temporarily to fulfill family responsibilities, to study, or for other reasons. Some workers leave permanently because they are interested in pursuing other occupations or because of dissatisfaction with hours, low pay and benefits, and stressful conditions.

Training, Other Qualifications, and Advancement

The training and qualifications required of childcare workers vary widely. Each State has its own licensing requirements that regulate caregiver training; these range from a high school diploma, to community college courses, to a college degree in child development or early-childhood education. Many States require continuing education for workers in this field. However, State requirements often are minimal. Childcare workers generally can obtain employment with a high school diploma and little or no experience. Local governments, private firms, and publicly funded programs may have more demanding training and education requirements. Some employers prefer to hire childcare workers with a nationally recognized childcare development credential, secondary or postsecondary courses in child development and early childhood education, or work experience in a childcare setting.

Other employers require their own specialized training. An increasing number of employers require an associate degree in early childhood education. Schools for nannies teach early childhood education, nutrition, and childcare. Childcare workers must anticipate and prevent problems, deal with disruptive children, provide fair but firm discipline, and be enthusiastic and constantly alert. They must communicate effectively with the children and their parents, as well as other teachers and childcare workers. Workers should be mature, patient, understanding, and articulate, and have energy and physical stamina. Skills in music, art, drama, and storytelling also are important. Self-employed childcare workers must have business sense and management abilities.

Opportunities for advancement are limited. However, as childcare workers gain experience, some may advance to supervisory or administrative positions in large childcare centers or preschools. Often, these positions require additional training, such as a bachelor's or master's degree. Other workers move on to work in resource and referral agencies, consulting with parents on available child services. A few workers become involved in policy or advocacy work related to childcare and early childhood education. With a bachelor's degree, workers may become preschool teachers or become certified to teach in public or private schools. Some workers set up their own childcare businesses.

Job Outlook

High replacement needs should create good job opportunities for childcare workers. Many childcare workers must be replaced each year as they leave the occupation to take other jobs, to meet family responsibilities, or for other reasons. Qualified persons who are interested in this work should have little trouble finding and keeping a job. Opportunities for nannies should be especially good, as many workers prefer not to work in other people's homes. Employment of childcare workers is projected to grow about as fast as the average for all occupations through the year 2012. The number of women of childbearing age (widely considered to be ages 15 to 44) in the labor force and the number of children under 5 years of age is expected to rise gradually over the projected 2002-12 period. Also, the proportion of youngsters enrolled full or part time in childcare and preschool programs is likely to continue to increase, spurring demand for additional childcare workers. Changes in perceptions of preprimary education may lead to increased public and private spending on childcare. If more parents believe that some experience in center-based care and preschool is beneficial to children, enrollment will increase.

Concern about the behavior of school-age children during nonschool hours should increase demand for before- and afterschool programs. In addition, the difficulty of finding suitable nannies or private household workers also may force many families to seek out alternative childcare arrangements in centers and family childcare programs. Government policy often favors increased funding of early childhood education programs, and that trend will probably continue. Government funding for before- and afterschool programs also is expected to be steady over the projection period. The growing availability of government-funded center-based care and preschool programs may induce some parents to enroll their children who otherwise would not do so. Some States also are increasing subsidization of the child daycare services industry in response to welfare reform legislation. This reform might cause some mothers to enter the workforce during the projection period as their welfare benefits are reduced or eliminated.

INDEX